TECHNOLOGY MARKETS AND
EXPORT CONTROLS IN THE

**DO NOT REMOVE
CARDS FROM POCKET**

ALLEN COUNTY PUBLIC LIBRARY

FORT WAYNE, INDIANA 46802

You may return this book to any agency, branch,
or bookmobile of the Allen County Public Library.

DEMCO

TECHNOLOGY MARKETS AND EXPORT CONTROLS IN THE 1990S

Also by David M. Kemme

Economic Transition in Eastern Europe and the Soviet Union: Issues and Strategies

Economic Reform in Poland (editor)

TECHNOLOGY MARKETS AND EXPORT CONTROLS IN THE 1990S

Edited by

David M. Kemme
Pew Economics Scholar-in-Residence
Institute for East-West Security Studies

NEW YORK UNIVERSITY PRESS
Washington Square, New York

Allen County Public Library
Ft. Wayne, Indiana

© 1991 by the Institute for East-West Security Studies, New York
All rights reserved
Printed in the United States of America

The Institute for East-West Security Studies does not take or encourage specific policy positions. It is committed to encouraging and facilitating the discussion of important issues of concern to East and West. The views expressed in this report do not necessarily reflect the opinions of the Board of Directors, the officers or the staff of the Institute.

Library of Congress Cataloging-in-Publication Information

Technology markets and export controls in the 1990s/edited by David M. Kemme.
    p.   cm.
  Includes bibliographical references and index.
  ISBN 0-8147-4617-9 : $30.00
    1. North Atlantic Treaty Organization. Coordinating Committee on Export Controls. 2. Export controls—International cooperation. 3. National security—International cooperation. 4. Technology transfer—Soviet Union. 5. Technology transfer—Europe, Eastern. 6. East-West trade (1945–) I. Kemme, David M.
HF1414.5.T43   1991   91-13602
382'.64—dc20   CIP

CONTENTS

Foreword	vii
Editor's Acknowledgments	ix
1. Introduction: Technology Controls and Prospects for Change *David M. Kemme*	1
2. Western Export Controls: An East European View *Andrzej Rudka*	17
3. Western Export Controls, International Technology Markets, and the Performance of the East European Economies *Volkhart Vincentz*	49
4. Global Technology Markets and Security Issues *Igor E. Artemiev*	73
5. US Export Controls in Transition: Implications of the New Security Environment *Gary K. Bertsch and Steven Elliott-Gower*	105
Colloquium Participants	129
About the Authors	133
Index	135

FOREWORD

As a new, post-Cold War security order develops in Europe, all elements of the old order are being closely reexamined. One such element is the system of restrictions on the export of certain Western high-technology items to Eastern Europe and the Soviet Union. If it is no longer in the West's interest to prevent the export of high-technology goods or information to the USSR and the newly reforming countries of Eastern Europe, what then is the most efficient and mutually beneficial approach to technology transfer? On September 14, 1990, the Institute for East-West Security Studies sponsored a day-long colloquium in New York on "Technology Transfer Issues in the 1990s" to discuss this topic.

Members of the academic, government, and business communities were invited to participate in the colloquium. Four papers were commissioned from experts in Poland, Germany, the Soviet Union, the United States, and the United Kingdom. These papers have been updated for this publication, and they represent the most current thinking of specialists in the East and the West on this crucial question. They are presented here with an introductory chapter by David M. Kemme, the Pew Economics Scholar-in-Residence at the IEWSS.

In this volume, the authors address the most pressing issues involving controls on high-technology exports: What has been the role of the Coordinating Committee for Multilateral Export Controls (COCOM), and what should its role be in the future? Have the negative consequences of technology control outweighed the positive? What role has central planning played in stunting technological innovation? Most importantly, what form should a new, post-Cold War export control regime take?

In the course of addressing these questions, the authors examine world technology markets; describe how trade restrictions on high technology have affected the civilian economies of East and West; and analyze the history, structure, and rationale of the export control systems of individual Western countries and of COCOM as a whole. Given the dramatic changes in the international security environment, the authors argue that new policies should be designed to enhance high-technology trade with the East while preserving legitimate controls on the transfer of sensitive military technologies to those countries still considered by Western leaders to be potential adversaries.

The Institute for East-West Security Studies would like to express special gratitude to the Pew Charitable Trusts, which provide leadership funding for the Institute's Economic Transition Program. The Institute would also like to thank the following philanthropic foundations for their generous support of IEWSS programs: The Ford Foundation, The Alfried Krupp von Bohlen und Halbach Foundation, The William and Flora Hewlett Foundation, The John D. and Catherine T. MacArthur Foundation, The McKnight Foundation, The Rockefeller Brothers Fund, The Sasakawa Peace Foundation, The Florence and John Schumann Foundation, and The Weyerhaeuser Foundation.

John Edwin Mroz
President
Institute for East-West
Security Studies

Anthony M. Solomon
Chairman
IEWSS Economic Transition Program

April 1991

EDITOR'S ACKNOWLEDGMENTS

There are many people to acknowledge in a joint effort such as this. First of all, I am grateful to the scholars who contributed to this volume and presented papers at the Institute's colloquium on "Technology Transfer Issues in the 1990s": Igor E. Artemiev, Gary K. Bertsch, Steven Elliott-Gower, Elina Kirichenko, Andrzej Rudka, and Volkhart Vincentz. Thanks are also due to the chairs and discussants at the colloquium: Janos Farkas, George Holliday, Sergei Kambalov, Marie Lavigne, Alexei Mozhin, Henry Nau, and Nikolaj Ordnung.

In addition, I would like to express my gratitude to Kristi Bahrenburg for an excellent job of organizing the colloquium and preparing the papers presented there; to Elizabeth Brainerd for research assistance; and to Richard Levitt, IEWSS Publications Director, for his helpful feedback on the manuscript as a whole. Finally, I would like to thank Rosalie Morales Kearns, IEWSS Publications Editor, for her work in copy editing and producing this volume and for her invaluable assistance in the research and writing of the introductory chapter.

TECHNOLOGY MARKETS AND EXPORT CONTROLS IN THE 1990S

1

Introduction: Technology Controls and Prospects for Change in the 1990s

DAVID M. KEMME

This collection, *Technology Markets and Export Controls in the 1990s*, presents four views on technology markets and controls in the next decade, prepared by scholars from Poland, Germany, the USSR, and (in a jointly authored chapter) the US and the UK. The authors describe the post-Cold War environment for technology transfer and make recommendations for implementing a new, lower-level export control regime. To set the scene for the next four chapters, I will briefly review the general functions and history of trade controls and particularly the export control systems developed by the US and the Coordinating Committee for Multilateral Export Controls (COCOM).

Over the past 40 years, widely varying views on the control of trade and technology transfer to Eastern Europe have been expressed; these views have ranged "from denial to detente."[1] Those who advocate complete denial of exports to the Soviet Union and former Warsaw Pact allies argue that any trade increases the efficiency of the Soviet economy, thereby releasing economic resources that may be

---

I would like to thank Gary K. Bertsch, Andrzej Rudka, Rosalie Morales Kearns, and Richard Levitt for helpful comments on an earlier draft of this paper.

1. Henry R. Nau, "Trade and Deterrence," *The National Interest*, no. 7 (Spring 1987).

utilized in military production. They see the denial of trade as part of a strategy of economic warfare, an integral part of the East-West conflict, where differences between the two socioeconomic systems are deemed irreconcilable. Detente and normalized trade relations, on the other hand, are seen as potential means of resolving the conflict. Proponents of detente see trade and business relations as a means of influencing Soviet behavior, and they believe that differences in socioeconomic systems may be reconciled with time and effort.

Actual policy toward Eastern Europe and the Soviet Union generally lies somewhere in between. Neither denial nor detente has been completely pursued or completely appropriate. There has been and is a need for selective trade controls. First, even in a period of detente, controls of strategic technologies are necessary for maintaining US and allied security, as long as there are nations that threaten that security. Second, selected trade restrictions or other economic sanctions may be used as an effective foreign policy instrument to manage the conflict between East and West. Variations in the level of controls or the imposition of sanctions are "signalling" devices that provide a foreign policy response more significant than rhetoric but less threatening than military options. They may also be utilized as a means of "disassociating" the US or the West from a particular event. Third, they may be used to limit Western dependence on East European and Soviet trade, limit the vulnerability of the West to Soviet intimidation, and therefore enhance Western economic security. Finally, trade restrictions and regulations may be used to extend the Western trading system (e.g., the General Agreement on Tariffs and Trade) and introduce market relations with the Soviet Union and Eastern Europe.[2]

Given that trade controls and economic sanctions can enhance security and serve as foreign policy instruments, the question that arises is, to what extent should controls be utilized, and when? Issues that have long been debated in the West include how to precisely define a strategic good, how to control re-exports and determine

---

2. For details and a policy-oriented discussion of these issues, see Nau, "Trade and Deterrence." For a similar review but in a more theoretical context, see Thomas A. Wolf, *US East-West Trade Policy* (Lexington, MA: Heath, 1973).

foreign availability of comparable goods, and how to solve jurisdictional disputes between nations.[3] A critical lesson of the history of trade controls is that the world of economics and politics is not static: a strategic good may become obsolete quickly, and the underlying political reasons for sanctions may readily disappear.

In a 1987 article, Henry Nau presents a cogent framework for addressing this problem.[4] East-West trade has long had three dimensions: strategic trade, foreign policy-related trade, and purely profit-oriented trade. Trade may be conducted at varying levels within a deterrence framework—defense plus dialogue—but constant review of trade controls and policies and close consultation with the allies are required. A strong COCOM becomes an essential element of such a system. Within this framework, what modifications in policy and institutions are necessary, now that several of the nations of Eastern Europe have moved so dramatically toward Western-style democracies with market economies?

Already, "more changes in the multilateral export control system were announced in June 1990, than during the entire period in which such controls have been used."[5] But recent developments necessitate an even more far-reaching review of export controls and the use of trade as an integral component of Western foreign policy toward the Soviet Union and Eastern Europe. The Warsaw Treaty Organization (WTO) and the Council for Mutual Economic Assistance (CMEA) have collapsed.[6] New governments, particularly in Poland, Hungary, and Czechoslovakia, are committed to introducing market-type economies. Modern technology, especially telecommunications equipment, is necessary for this transformation. As a result of these developments,

---

3. For a broad survey of these issues, see Gary K. Bertsch, ed., *Controlling East-West Trade and Technology Transfer: Power, Politics and Policies* (Durham and London: Duke University Press, 1988); and Adlai E. Stevenson and Alton Frye, "Trading With the Communists," *Foreign Affairs* 68, no. 2 (Spring 1989), pp. 53-71.

4. Nau, "Trade and Deterrence."

5. Glennon J. Harrison, *The Bush Administration's New Approach to Export Controls and COCOM*, CRS Report for Congress, August 8, 1990.

6. On April 1, 1991, the WTO ceased to exist as a military organization, and the CMEA is expected to be formally dismantled in late June, 1991.

many have challenged the rationale for continued export controls of certain necessary technologies, especially because the success of the current market-oriented and democratizing reforms may be at stake.

It is quite difficult, however, to ascertain the extent to which economic growth in the Soviet Union and Eastern Europe was hindered by Western export controls, and therefore, to what extent relaxation of trade restrictions will stimulate growth and development. The nature of the centrally planned economy, in which the process of economic decision making was highly politicized, was also an important factor in preventing full Soviet and East European participation in the world market. As Josef Brada observed, "studies of Soviet technological progress during the 1950s and early 1960s suggested that technological progress in the Soviet Union was somewhat faster than in the United States. . . . Since then, studies . . . show a sharp decline in the growth of technological progress."[7] In the 1980s, the levels of competitiveness and technological progress in the socialist economies lagged even further. This decline was caused in large part by domestic or systemic inefficiencies, but also by misguided investments and economic policies such as those that guaranteed full employment, the lack of an incentive system to encourage innovation in enterprises, and restrictions on travel, which limited scholars' access to foreign technology and scientific exchange.

Thus, the socialist countries were made even more vulnerable to Western trade controls by domestic systemic constraints and policy decisions. While it may be argued that restrictions on high-technology trade were indeed an effective tool of international policy in the long run (given the events of the last two years), it also appears that relaxation of certain trade controls will have little immediate impact. The full benefits of more technologically advanced communications systems, for example, will take years to materialize. And a simple relaxation of controls will not be sufficient to ensure transition to market economies in the East.

Before turning to the chapters below, in which the authors take up

---

7. For several case studies, see Bruce Parrott, ed., *Trade, Technology, and Soviet-American Relations* (Bloomington, IN: Indiana University Press, 1985). For a broad overview, see the chapter by Josef Brada, "Soviet-Western Trade and Technology Transfer: An Economic Overview," in that volume.

these and related issues, it may be useful to provide a brief overview of the system of Western export controls, which consists of COCOM and the individual national control systems. Here I will focus on the history of COCOM and the legislation and policies of the US system, since that system plays an integral role in COCOM and has been the origin of many disputes among the allies.

US Export Controls and COCOM

The Trading With the Enemy Act, passed in the US in 1917, was the precursor to the current system of export controls. Under this act, the US president was authorized to prohibit economic activity in wartime with states designated as enemies. During World War II, restrictions on exports formed part of the US war strategy. In the late 1940s, the US targeted the Soviet Union and China as proscribed destinations; and the Export Control Act of 1949 served to codify what has been referred to as "the economic equivalent of political 'containment.' "[8] A parallel system was initiated in England and France in 1949. It was soon realized that effective control of exports to the Soviet bloc required multilateral coordination. At the initiative of the US, a voluntary Coordinating Committee was established in 1949. The original COCOM lists were based on the British and French lists, but the US Battle Act of 1951 provided the means of ensuring the participation of Western allies.[9] In the early years of the Cold War, there was little problem with persuading the allies to comply with COCOM regulations—the Berlin Wall and the Korean War were sufficient persuasion.[10]

---

8. Benjamin J. Cohen, quoted in Wolf, *US East-West Trade Policy*, p. 48.

9. "Although members had no legal obligation to abide by it, the Battle Act of 1951 stipulated that no military, economic, or financial assistance would be supplied to any nation unless it embargoed shipments to any nation . . . threatening the security of the United States." Gary K. Bertsch, *East-West Strategic Trade, COCOM and the Atlantic Alliance* (Paris: The Atlantic Institute for International Affairs, 1983), p. 14.

10. William A. Root, "Trade Controls that Work," *Foreign Policy*, no. 56 (Fall 1984), pp. 61-80. There was, of course, significant ongoing debate on the scope of controls. On this see also Michael Mastanduno, "The Management of Alliance Export Control

The Western system of export controls continues in the form of parallel country systems, which are coordinated by the informal (non-treaty), multinational COCOM, whose decisions are codified in the domestic legislation of each member.[11] The enforcement of trade controls is the responsibility of individual nations, while COCOM remains a forum for negotiations and consultations among allies, and a means of adjudicating requests for exceptions to certain controls.[12] There is no supranational authority to police or enforce the COCOM lists, although there are other bodies that monitor certain trade flows.[13]

COCOM maintains three lists: two of them, the Munitions List and the Atomic Energy List, are fairly clearcut, and there is ready agreement that the items on these lists qualify as strategic. The International Industrial List, however, is the source of controversy, as it regulates dual-use technology, that is, high-technology industrial items with potential military applications. If a product is included on one of the control lists, it is not necessarily automatically banned from export, since an exceptions procedure may be invoked.

In 1954 and 1958, the COCOM embargo lists were reduced significantly. Throughout the 1950s, the general pattern of East-West trade remained fairly constant, although in some cases the US applied different policies to different countries, for example, extending more

---

Policy: American Leadership and the Politics of COCOM," in *Controlling East-West Trade and Technology Transfer*, ed. Bertsch.

11. COCOM's original members were the United States, the United Kingdom, France, Italy, the Netherlands, Belgium, and Luxembourg, who were later joined by Norway, Denmark, Canada, the Federal Republic of Germany, Portugal, Japan, Greece, and Turkey. During the 1980s, COCOM expanded to include Spain and Australia. Switzerland, Sweden, Austria, Finland, South Korea, and Singapore adopted comparable trade regimens.

12. The essays in *Controlling East-West Trade and Technology Transfer*, ed. Bertsch, include examinations of the export control systems in Japan, the FRG, France, the UK, and the US. For a further examination of the export control systems in France, the FRG, and the US, see also Angela E. Stent, *Technology Transfer to the Soviet Union* (Bonn: Forschungsinstitut der Deutschen Gesellschaft für Auswärtige Politik e.V., 1983).

13. For a review of the objectives of COCOM, see William A. Root, "COCOM: An Appraisal of Objectives and Needed Reforms," in *Controlling East-West Trade and Technology Transfer*, ed. Bertsch.

liberal treatment to Poland and much harsher treatment to Cuba. Throughout the 1960s, there was relative harmony among COCOM participants. Although the number of items on the control lists was reduced, national security objectives were still well served, as the effectiveness of controls is hardly a function of the number of commodities on the lists.[14] COCOM controls were never intended to stop all trade, only certain types of militarily relevant trade.

During the era of detente, controls were gradually further relaxed, and trade with the East grew. In the United States, the Export Control Act of 1949 was replaced with the more pro-trade Export Administration Act of 1969, which solicited the input of business and commercial interests through Technical Advisory Committees for industries subject to export controls. The US shortened its list of controlled goods and technologies, removing items that were either insignificant to adversaries' military capacity or that were accessible on the international market. Nonetheless, the industrialized West's share of high-technology exports to the Soviet Union was 5% in 1970; this peaked at 7% in 1975.[15] While trade grew modestly, the West's policies appeared to have little impact on the overall level of military and economic development of the Warsaw Pact nations.

During the 1970s, as East-West trade increased, the number of exception cases also increased—from 228 in 1966 to 544 in 1970, 1,798 in 1975, and 1,680 in 1978. The US share of COCOM exception requests rose from 12.7% of the total in 1966 to 62.5% of the total in 1978.[16] This trend was the result of the fact that the pace of political and technological change outstripped the pace of changes in the COCOM list. Many exception requests from US manufacturers were granted because they met the "foreign availability" criterion provided in the Export Administration Act of 1969. If the technology was readily available from foreign sources, an export license for US manufacturers was relatively difficult to deny. Some argue that this resulted in a

---

14. For a discussion of measuring the effectiveness of COCOM, see Mastanduno, "Management of Alliance Export Control Policy."

15. Bertsch, *East-West Strategic Trade, COCOM and the Atlantic Alliance*, p. 16.

16. Mastanduno, "Management of Alliance Export Control Policy."

significant reduction in COCOM's effectiveness during the 1970s, as high technology leaked to the Soviet Union.

While exceptions increased, there were also notable examples of covert sales and even overt circumvention, which occurs when a COCOM participant government licenses a sale despite an objection in COCOM.[17] The number of cases of overt circumvention and covert sales appeared to have increased through the 1970s and into the 1980s, and was even seen as a threat to the functioning of the multilateral control system.

The US Export Administration Act of 1979 continued its predecessor act's focus on expanding exports, although it did not challenge the underlying rationale for controls or diminish the emphasis on protecting national security. The new act was intended to continue to expand trade, but in fact, the early 1980s saw a measurable increase in the restrictiveness of export controls (as a result of US policy in response to the invasion of Afghanistan and the imposition of martial law in Poland), and their use as a policy instrument became perhaps haphazard. Prior to 1978, nearly all COCOM disputes had centered around technical issues, questions of whether or not particular goods or technologies were strategic, issues concerning exception cases, and so forth. In 1978, the Carter administration utilized export controls as foreign policy tools, especially in response to human rights violations. Controls were introduced on gas and oil exploration and production equipment, as a reaction to the conviction of Soviet dissident Anatolii Shcharansky and the arrest of a US businessman. Sales of this type of equipment, which would likely have gone to US firms, immediately went to a French company instead.[18] After the Soviet invasion of Afghanistan in December 1979, all US licenses were suspended and no new ones issued, but by mid-1980 this decision was reversed. Then, in response to the December 13, 1981 declaration of martial law in Poland, the US reimposed controls on oil and gas exploration and production equipment, and transmission and refining equipment as

---

17. This was not a completely new phenomenon. In *East West Strategic Trade, COCOM and the Atlantic Alliance* (p. 40), Bertsch notes the case of the British sale of Viscount aircraft to China in the 1960s as an example.

18. Root, "Trade Controls that Work," pp. 64-65.

well. This was the second reversal of the general policy of encouraging Soviet oil and gas exports to the West (which had prevailed from 1977 to 1981), and led to the gas pipeline dispute between the allies, which lasted a year and a half.

In June 1982, the response to Polish martial law was hardened as the US expanded the controls to include foreign-origin goods of US subsidiaries and exports of foreign-produced goods using licensed US technology. The allies rejected the extraterritorial reach of US law, and the US responded by denying exports to allied firms violating these restrictions. The issue was finally resolved in November 1982, when President Reagan suspended the unilateral controls of December 1981 and June 1982. In December 1982, the allies agreed to undertake "studies intended to harmonize allied policies on East-West economic relations."[19] The dispute had little to do with whether the goods in question were strategic; rather, it focused on the question of whether that particular set of controls should have been implemented at the time for foreign policy purposes. In fact, during regular COCOM negotiations in 1980, rather tough controls had been agreed upon for well-defined strategic products. Truly strategic export controls were not endangered.

What became clear during the early 1980s was that, first, COCOM itself was effective in setting strict levels of export controls when the normal procedures of consultation and negotiation were employed. Second, attempts by the US to unilaterally employ trade controls as an instrument of foreign policy were highly controversial and only marginally effective.[20] Third, US efforts to pressure allies into similar actions generally failed completely. The intent of both the US and the West Europeans was unclear, and the true concerns of each side were not on the table.[21] A more open process of consultation could have alleviated some differences and left the remaining disputes

---

19. Ibid.

20. While they had virtually no real economic impact, they did serve the function of a signalling device. Better methods to achieve this goal could likely have been devised if this was the only objective, however.

21. See Heinrich Vogel, "Technology for the East—the Tiresome Issue," *Aussen Politik* 36, no. 2 (1985), pp. 117-26

at least clear and understandable. Perhaps a more active COCOM process could have alleviated the allies' dispute.

During the 1980s, the Reagan administration continued to express concern about the flow of Western technology to Soviet military efforts. Stricter licensing procedures were introduced in the US, enforcement efforts were increased, and the Defense Department took on a more prominent role in approving licenses for high-technology exports to certain allies as well as to communist countries. However, domestic critics of US policy began to emphasize the impact on US competitiveness in world markets, claiming that the government was controlling many items that had little if any military significance, that other items were readily available on world markets, and that the licensing process was slow and cumbersome.

The Omnibus Trade and Competitiveness Act of 1988 attempted to address these concerns through a series of amendments aimed at ameliorating export disincentives without a simultaneous reduction in the efficacy of the control regime. Progress in implementing these amendments had not yet been made when pressures from both US allies and US business induced the Bush administration in early 1989 to remove some controls originally implemented for foreign policy reasons. In May 1989, the "no exceptions" policy restricting high-technology exports to the Soviet Union (introduced in response to the Soviet invasion of Afghanistan) was removed, and in July 1989 controls on older but still rather sophisticated personal computers were removed. These steps were taken prior to any significant change in Eastern Europe. While the events in Eastern Europe were unfolding, but independently of them, the Bush administration removed the last export sanctions imposed by President Carter in response to the Afghanistan invasion. Export controls on equipment to the Kama River and ZIL truck plants were removed in February 1990. And, more importantly, in May 1990 the administration completed an internal review of export controls and COCOM and had developed major proposals for reform to present to the allies.

The Bush administration proposed a leaner "core" list that concentrated on more exclusively strategic goods; it recommended partial or complete relaxation of controls on more than one-third of the previously proscribed items (47 out of a total of 120). The proposed changes also included provision for easing restrictions on certain technology exports to the Soviet Union to the "PRC Green Line," i.e., a technology level of exports to the People's Republic of China that required nothing more than formal notification to COCOM. The political strides made in Eastern Europe were recognized through proposals to ease export licensing requirements for those countries in the region that adopted safeguards ensuring against diversion of controlled goods to proscribed destinations or uses.[22]

A June 1990 high-level COCOM meeting produced a decision to develop a new core list of controlled goods and technologies, constructed from scratch and based on eight categories: telecommunications; computers; navigation and avionics systems; propulsion systems; sensors, sensor systems, and lasers; electronics design, development, and production; advanced materials and material processing; and marine technology.[23] In addition, the COCOM members reached specific agreements on revisions in controls for three categories: machine tools, telecommunications, and computers. The members also agreed to extend favorable exception request consideration to those East European countries posing the least strategic risk to the West—most likely Poland, Hungary, and the CSFR. In addition to supporting COCOM modifications, the US administration moved to suspend the 1974 Jackson-Vanik Amendment, which linked trade to Jewish emigration from the USSR.

While the results of the June 1990 COCOM meeting marked tremendous change in export control regimes, continued progress in list revisions and US initiatives by the Bush administration and in Congress have slowed, due to the fluidity of Soviet and East European domestic politics. Recent events in the Soviet Union—among them the republic-union debate on the new union treaty and the crackdown in the Baltics in January 1991—have created a new atmosphere of caution. However, it is unlikely that these events will alter the fundamental structure of strategic trade controls.

What we have seen in the last year and a half is that COCOM as an organization has proven to be an effective means of coordinating trade

---

22. Harrison, *The Bush Administration's New Approach to Export Controls and COCOM*, p. 4.

23. Ibid.

control policy and negotiating differences among allies. A prudent approach of deterrence is still appropriate, and, given the increasing domestic instability in the region, even more caution may be warranted in the near term. While broad security objectives—e.g., protecting a lead in strategic technologies—remain the same, defining strategic goods remains difficult, and now the proscribed destinations may be modified. The new issues addressed below include the following: How can trade control regimes be modified in ways that meet all the West's security objectives while at the same time enhancing both the economic transition and the democratization process in the reforming East European countries? What new organizations or informal groups should be formed to control proliferation in other areas, and which countries should participate? Is existing COCOM membership sufficient, or should its membership be expanded to include the transition economies, and under what conditions? What criteria should then be employed to define the proscribed destinations?[24]

The Role of Export Controls in the 1990s

A subject of ongoing debate is the possible detrimental effects suffered by the economies of Eastern Europe and the Soviet Union as a result of technology controls. On one hand, some observers believe that Western technology has been the determining factor in much of the Soviet Union's technological and military progress, whereas others argue that the economic conditions of the importing country determine how far new technologies can be efficiently absorbed and diffused. Some also claim that imported technology may have actually harmed the Soviet economy, because the utilization of foreign technology allowed the postponement of economic reform. The contributors to this volume focus on different aspects of this issue, but they are unanimous on one point: the need for reassessing and revising the current system of Western export controls.

---

24. As noted by Bertsch and Elliott-Gower in this volume, export controls have already been used as foreign policy instruments concerning various issues (human rights, nuclear non-proliferation, etc.) against a number of countries.

Technology Controls and Prospects for Change in the 1990s 13

In chapter 2 of this volume, "Western Export Controls: An East European View," Andrzej Rudka provides an overview of COCOM's structure and the evolution of the Western system of export controls in the 1980s. He acknowledges the difficulty of quantifying the effect of controls as compared to the effects of domestic economic policies in the affected countries, but he does argue that certain conclusions can be drawn, i.e., that trade restrictions have allowed the West to maintain its technological lead over the East, and have exacerbated the East European countries' economic problems, particularly in nonmilitary sectors, "by forcing them to rely on . . . less efficient technology . . . and by forcing them to significantly delay the realization of certain economic goals."

Rudka also argues that recent reforms in Central and Eastern Europe have undermined the rationale for Western restrictions on exports to certain destinations in the East. He makes specific short-term recommendations and more general long-term recommendations for reforming export regulations and for setting up a technology safeguards system, in which Central and East European countries could sign agreements to prevent the diversion of COCOM-controlled items to the remaining proscribed destinations. Each country's final goal would be to create a safeguards system in such close cooperation with Western export controls that the country would be removed from the list of COCOM proscribed destinations.

Volkhart Vincentz, in chapter 3, "Western Export Controls, International Technology Markets, and the Performance of the East European Economies," also argues that, while domestic conditions contributed to the low level of technology and competitiveness in Eastern Europe, Western export controls were also an important factor. Export controls on basic high-tech components prevented Eastern Europe from producing competitive exports and specializing in low-tech assembly of high-tech products. Another result of export controls was that East European countries were forced to devote resources to domestic production of technology, with disappointing results. While some contend that in Eastern Europe there is low demand for high technology, which anyway cannot be absorbed and utilized efficiently, Vincentz points out that some technologies have proliferated quickly, and still others (such as personal computers) would be utilized if the necessary technological infrastructure were in place. In any case, now that substantial reforms have begun, the transformation to a market economy will eliminate the economic mechanisms that had hindered diffusion. In addition, the reforms have created even greater demand for high technology (e.g., telecommunications for setting up money and capital markets, and technology for environmental cleanup).

Vincentz also discusses characteristics of international technology markets that could make it difficult for East European companies to compete. Western firms have been able to take advantage of economies of scale, cooperation on R&D projects, exchange of technological know-how, and the large "home markets" of multinational firms. He concludes his chapter with a warning against tendencies toward interventionist trade policies in both East and West; and he calls for the West to open its markets to East European products and to remove barriers to intra-industry trade.

Igor E. Artemiev describes the major characteristics of world technology markets and provides a detailed analysis of the noncommercial nature of technology diffusion in a command economy in chapter 4, "Global Technology Markets and Security Issues." In command economies, technology was seen as a public good and was financed by the state. It was "collectivized," because ownership of technology was considered a hindrance to diffusion. The inventor of a new technology was recognized by the "author's certificate" and received some royalties, but the state retained all rights of use. Enterprises lacked incentive to transfer technology, because they could not recoup their investment costs, in part because they had to diffuse the innovation without compensation. This noncommercial type of technology diffusion was used within and between CMEA member states, and between CMEA member states and less developed countries. In transactions with market economies, however, technological know-how was treated as a commodity. Sales were based on market prices and carried out through the foreign trade organization that held the monopoly on imports and exports of disembodied technology.

The late 1980s saw a liberalization of state control over technology transfer. Joint ventures with foreign firms have been legalized, domestic firms can now handle their own import and export transactions directly, and new legislation has restored the use of patents. Nevertheless, there is still a need for "a fundamental reform of the Soviet economy and the creation of product, technology, capital, and labor markets" in order to overcome the USSR's technology gap with the West.

Artemiev expects that export controls will remain in place, at least for the foreseeable future, because Western leaders are still not convinced that the Soviet Union no longer poses a military or economic threat to the West. He advocates a number of confidence-building measures, within the existing framework of controls, that might lead to the easing of restrictions in the future. Among these measures are a formalized structure for East-West discussion on export controls, reduction of items on the COCOM lists, more transparency of export regulations, the creation of verification regimes for "potential adversaries," and an ad hoc panel of experts from East and West to study export restrictions.

Artemiev suggests that verification techniques developed in the context of arms control negotiations could serve as a valuable example for an export control verification system that would provide for on-site inspections of end-users in the importing countries. The possibility of establishing "parameters of acceptable uncertainty" is another concept that should be borrowed from the military sphere.

According to Artemiev, even a shortened COCOM list is an inappropriate remnant of Cold War thinking that damages trust and seriously hinders East-West economic cooperation. Confidence-building measures must be used to dismantle what Artemiev refers to as one of the remaining walls separating East from West.

In chapter 5, "US Export Controls in Transition: Implications of the New Security Environment," Gary K. Bertsch and Steven Elliott-Gower discuss the need for a reassessment of US export control policy in light of the new security environment, which renders Cold War concepts of security obsolete. The most essential aspects of the new security environment include the end of the Cold War, the USSR's continued status as a superpower, the necessity of Western assistance to the current reforms in Eastern Europe, the importance of economic and technological considerations for a new concept of security, the relative decline in US economic power and technological superiority, and the proliferation of weapons of mass destruction throughout the world.

These developments have numerous and far-reaching implications for US export control policy. COCOM restrictions have been significantly eased, and talks are under way with certain Central European countries to establish technology safeguards systems. Although US policy makers are not yet ready to dismantle export controls altogether, more and more of them are subjecting these controls to a rigorous cost-benefit analysis. In the face of a growing number of alternative suppliers of high technology, it will be increasingly difficult for the US to maintain export controls on certain products. East-West cooperation on export controls must be developed in order to combat the proliferation of weapons of mass destruction. Bertsch and Elliott-Gower stress the importance of a system that is more inclusive than COCOM, "incorporating both the NICs and the West's former communist adversaries," to achieve that end, in the framework of a reformed COCOM, the CSCE, or some sort of treaty on the model of the Nuclear Non-Proliferation Treaty, to be coordinated by the United Nations.

Conclusion

The transition to a market economy and the import of high technology are closely linked. The latter remains conditional, and possible only when progress in reforming the political and economic systems is clearly evident. Recent events indicate that progress in the transition will be slow and painful. Political instability within the Soviet Union calls for caution, and each country in Eastern Europe must be considered on a case-by-case basis. It is likely, however, that the 1990s will see significant changes in the Western export controls system. In fact, these changes are critical to the fulfillment of the promises of the revolutions of 1989 and 1990.

2

Western Export Controls: An East European View

ANDRZEJ RUDKA

This chapter examines the Western system of export controls from the East European standpoint, i.e., from the standpoint of the countries that have been the targets of these trade restrictions for 40 years. I will examine the system's main characteristics, as well as its strengths and its weaknesses, which are especially visible in the present transitional period in East-West political and economic relations. The new security environment has precipitated changes in the attitudes of the US and of the West in general, and movement is being made toward significant adjustments in the system. I will discuss long-term and short-term policies by which the Central and East European countries can contribute to those changes and make them irreversible. This could result in a decisive refocusing of the Western export control system and the transformation of at least some of its previous targets into cooperating or even full members of the system.

Main Characteristics of the Export Control System

The system of Western restrictions of technology transfer to the East consists of two overlapping parts: the Coordinating Committee for

Multilateral Export Controls (COCOM) and national control systems in individual member countries. COCOM was created in late 1949, at the beginning of the Cold War, as a result of intensive political activity by the United States and other West European countries, particularly France, the UK, and Italy. The intent was to multilateralize the export restrictions that had been introduced individually by those countries in late 1947.

COCOM is an informal international organization whose members include the majority of developed market economies,[1] and which cooperates indirectly with a group of other countries.[2] It sets guidelines and coordinates the West's policy of export controls towards the East.[3] The vast majority of member countries follow the guidelines set by COCOM on the scope of export controls; only a few COCOM participating countries, including the United States, have export control systems that are organizationally larger or that cover a broader scope of internationally traded commodities than COCOM. The

---

1. Belgium, France, the United Kingdom, Italy, Luxembourg, Netherlands, and the United States, which took part in negotiations on the creation of such an organization finalized by an agreement signed in November of 1949, became the first members of COCOM. In 1950, they were joined by Canada, Denmark, Norway, Portugal, and the Federal Republic of Germany, in 1952 by Japan, and in 1953 by Greece and Turkey. Spain entered COCOM in 1985 and Australia in 1989. Altogether, COCOM consists of 17 countries—all the NATO members (except Iceland) plus Japan and Australia.

2. In the early 1950s, over 50 non-COCOM countries cooperated with the United States in enforcing export control measures. In the 1980s, the US put very strong, and largely successful, pressure on West European neutral countries and the newly industrialized countries of Southeast Asia, and generally gained their acceptance of COCOM's control rules. See Henry R. Nau, *The Myth of America's Decline: Leading the World Economy into the 1990s* (New York: Oxford University Press, 1990). As a result of that, close to 15 strategic trade agreements have been signed so far by the United States with non-COCOM countries, in which the latter agreed to comply with multilateral export restrictions (this group includes Austria, Finland, Sweden, and Switzerland, as well as India, Singapore, South Korea, and Taiwan).

3. The geographical scope of COCOM export controls covers the Central and East European countries, i.e., Albania, Bulgaria, Czechoslovakia, Hungary, Poland, Romania, and the Soviet Union, and a few other socialist-oriented countries of Asia, e.g., Afghanistan, China, Cambodia, Laos, Mongolia, North Korea, and Vietnam. The controls also applied to the German Democratic Republic before German unification.

United States, long the unquestionable world leader in technology, has always played a decisive role in shaping the direction of COCOM's development.

It is worth underlining the informal character of COCOM. This organization is not based on any international treaty. It was created in 1949 and has functioned since that time on the basis of an informal understanding on the part of the participating countries. Such a legal solution was chosen by negotiators mainly because the introduction of such controls encountered serious resistance in many countries (mainly European ones), from parliaments and the general public, which was largely sympathetic to the East in that period. In view of its informal status, one may regard COCOM as simply a substitute for a political (strategic) export cartel of the participating countries.

The informal character of COCOM has certain essential consequences for its functioning and for the effectiveness of its controls. Participating countries are not legally bound by the (usually) unanimous decisions of COCOM members (or at least of representatives present at a session). Thus, those decisions are more like recommendations that should be implemented within the national legal and export control systems. There is also no formal supranational mechanism to supervise the implementation of those decisions by individual countries, and no mechanism to discipline member countries for not complying with the rules agreed upon. Therefore, the United States has often tried to directly and unilaterally influence the activities of other countries in a direction consistent with its own goals, sometimes even invoking extraterritorial use of its own export control regulations.[4] The lack of openness in COCOM's functioning, largely a result of the substance of its activities, is to some extent also a consequence of its informal status.

It is worth noting, however, that international law has no direct bans or even clear-cut provisions on export restrictions such as those

---

4. The most recent case took place in 1982, when the United States extended its restrictions on exporting pipeline equipment to the Soviet Union to cover US affiliates abroad (US subsidiaries and licensing affiliates). However, the West European allies rejected this invasion of their sovereignty, and the British, French, Italian, and West German governments invoked national blocking legislation that protected their companies from US legislation. See Nau, *Myth of America's Decline*.

discussed here, which discriminate against particular countries in international trade turnover. One can find certain guidance in that respect only from general standards and rules of behavior in international relations, e.g., the principle of peaceful coexistence of all countries, broadly accepted by all UN members, and other principles ensuing from that rule, especially those of proper international cooperation, nonintervention in other countries' affairs, nonaggression, and rejection of the use of force in mutual relations among sovereign countries.

This being said, one may conclude that international law indirectly imposes on international actors a commitment to refrain from trade discrimination motivated by noneconomic factors. However, article 21 of the General Agreement on Tariffs and Trade (GATT) allows individual countries to apply measures necessary—according to their own judgments—to protect their national security interests. Thus, Western export controls to Eastern Europe are de facto sanctioned by international law, but only in the case of individual countries; there is no provision for an international organization especially created for that purpose.[5]

Therefore, it is quite probable that COCOM will continue to exist in its present status, irrespective of many recurring attempts— undertaken mainly by the United States (most recently in the early 1980s)—to change it into a formal organization. However, recent COCOM meetings have called for all participating countries to make their technology export controls better known, better understood, and therefore more widely accepted by the general public.[6]

The principal objective of COCOM is to control the export of proscribed commodities to designated countries. *Proscribed* generally means *strategic*, i.e., capable of strengthening military potential. However, to define a strategic commodity in a way that is acceptable to

---

5. This may be another good reason that COCOM has been kept informal for so many years.

6. From time to time it has even been argued by some Western experts that formalizing COCOM and transforming it into a full-fledged international organization may be a positive development for the proscribed countries as well. It would make the whole export control system more transparent for them and, therefore, more predictable.

all COCOM member countries is a problem that has never been fully resolved. Since the mid-1970s, the West's rationale for applying export controls has usually been formulated as the need to keep its technological lead over its adversaries, i.e., the socialist countries, with the Soviet Union in the forefront. This strategic goal was officially adopted by COCOM member countries in the mid-1980s.

Irrespective of its declared strategic character, the COCOM system of multilateral export controls has been frequently used for other than strictly strategic goals, depending on the actual state of East-West political tensions. It has often been used to further purely political goals, as was the case in the early 1980s.[7] On the other hand, during periods of lessened East-West political tensions, restrictions were usually eased (most significantly in the 1970s for most proscribed countries and in the mid-1980s for China). Nevertheless, it is worth noting the persistent asymmetry in the pace of tightening and relaxing export controls. Usually export controls have been tightened very quickly in response to any serious "misbehavior" on the East European side, while the relaxation of controls had to be thoroughly considered, multilaterally discussed, and carefully balanced.[8]

The commodity scope of export controls is formulated on three principal COCOM lists. Two of them—the Munitions List and the Atomic Energy List—include clearly defined military-capable and strategic items: controls of exports on these lists do not raise any serious objections on the part of COCOM member countries and are generally well understood on the East European side. The International Industrial List constitutes the main source of problems and the principal objective of most of COCOM's workload. It covers so-called dual-use entries, i.e., technologies with both civil and military applications.

---

7. This manifested itself, for example, in the form of a sudden increase in commodities subject to control and in the strengthening of the regime of export controls in general and with regard to individual countries in particular (in the early 1980s this concerned mainly the Soviet Union and Poland).

8. The experience of the early 1980s, as well as present discussions within COCOM concerning relaxation of controls, fully confirms that observation. See Kevin F.F. Quigley and William F. Long, "Export Controls: Moving Beyond Economic Containment," *World Policy Journal* 7, no. 1 (Winter 1989-1990).

The COCOM lists have been, in principle, secret.[9] Nevertheless, their general contents are included in the member states' national lists of controlled items, since these are based on multilaterally accepted COCOM lists. In the mid-1980s, the International Industrial List consisted of about 150 entries (compared to around 130 in 1980), while the Commodity Control List of the US had over 240 entries (215 in 1980), that of Japan—over 180 entries, of France—over 150 entries, and of the United Kingdom—123 entries. This indicates an overall increase in the number of items on the lists in comparison with the late 1970s, though since the mid-1980s a trend to limit their scope has again been noticed. As of the late spring of 1990, there were 116 entries on the COCOM International Industrial List.[10]

Nevertheless, considering only the number of entries on the lists is not enough to assess the actual level of restrictiveness of export controls. On the one hand, estimates from the late 1980s put the real number of internationally traded goods covered by export controls at around 2,000 individual commodities.[11] On the other hand, there are mechanisms in COCOM for relaxing the formal regime of export controls by allowing the export of commodities covered by the lists to proscribed destinations on a case-by-case basis. This "exception request" principle may be applied under specific conditions, including importers' commitments to comply with the requirement not to allow an unauthorized use of acquired items or the transshipment of those items to other proscribed destinations.

---

9. In 1985, the UK published a general outline of its lists in "Security Export Controls," *British Business* (Supplement), 1985. There was also information that the full COCOM industrial list (excluding classified information) would be published for the first time in the United States in the near future. "Some Must Reading on High-Tech Sales," *International Herald Tribune*, June 23-24, 1990.

10. "US to Relax Standards on High-Tech Exports," *The New York Times*, June 8, 1990. According to other sources, there were 120 items on the list (see, for example, Glennon J. Harrison and George Holliday, *Export Controls*, CRS Issue Brief, Updated June 21, 1990).

11. According to another source, "the number of controlled products is so vast that the Department of Commerce itself cannot provide a precise count, though the unofficial estimate is in the hundreds of thousands." See Quigley and Long, "Export Controls."

To fully assess Western export controls, we must consider their impact on East-West trade and the economies of the countries involved; we must also examine the effectiveness of those restrictions, that is, how well they have achieved their primary goals.

Numerous analyses show that export controls have always significantly influenced East-West trade patterns. In particular, controls have reduced global exports of technologically advanced industrial commodities to Eastern Europe, thus distorting both its commodity structure and its geographic pattern of trade, as the East European import demand in those areas has been continually shifted to other countries that apply a less restrictive regime of controls or do not apply them at all.[12] This is true for all periods, i.e., the 1960s and 1970s as well as the 1980s, when Western export controls were severely strengthened and technology transfer to the East (except China) was significantly reduced.

There is also no doubt that Western export controls have imposed essential, though practically impossible to calculate, costs on the civilian economies of proscribed countries by forcing them to rely on less technically sophisticated, and thus less efficient, technology in many branches of industry and services (e.g., telecommunications) and by forcing them to significantly delay the realization of certain economic goals (e.g., computerization of the banking system, the lack of which is nowadays—at least in Poland—one of the critical constraints on the functioning of enterprises under the new economic conditions). This is true even if one argues that the more binding constraints of the economic development of Eastern Europe are

---

12. See, among others, Josef C. Brada and Larry J. Wipf, "The Impact of US Trade Controls on Exports to the Soviet Bloc," *Southern Economic Journal* 41, no. 1 (1974), pp. 47-56; Thomas A. Wolf, *The Quantitative Impact of Liberalization of US Unilateral Restrictions on Trade with the Socialist Countries of Eastern Europe* (Washington, DC: Department of State, 1972); Thomas A. Wolf, "The Impact of Formal Western Restraints on East-West Trade: An Assessment of Existing Quantitative Research," in *Tariff, Legal and Credit Constraints on East-West Commercial Relations*, ed. John P. Hardt (Ottawa: Carleton University, 1975); and Andrzej Rudka, *Wpływ polityki kontroli eksportu USA na transfer zachodniej technologii do krajów socjalistycznych w latach osiemdziesiatych* [The impact of US export controls on Western technology transfer to socialist countries in the 1980s] (Warsaw: Instytut Koniunktur i Cen Handlu Zagranicznego [Foreign Trade Research Institute], 1982).

probably domestic in origin, especially those elements of the economic system that reduced or removed the incentives for East European enterprises to innovate. Unfortunately, it is not possible to calculate the separate impact of those and other plausible causes of the poor economic performance of Central and East European countries.[13]

At the same time, however, substantial costs have also been incurred by the countries that apply export controls. A 1987 report of the National Academies of Sciences and Engineering and the Institute of Medicine estimated the direct costs to the US economy at $9.3 billion in 1985 alone, which also meant a loss of approximately 188,000 jobs. The total annual costs (direct and indirect, i.e., accumulated in the longer term) were put at $17.1 billion.[14]

All this having been said, the question of the effectiveness of export controls still remains to be answered. There have been some spectacular failures to introduce embargoes on certain technology flows to the East.[15] But we cannot judge Western export controls completely ineffective simply on the basis of these failures. Usually those restrictions were politically motivated sanctions rather than

---

13. See Economic Commission for Europe, *Economic Reform in the East: A Conceptual Framework for Western Support* (Geneva, 1990). If we accept earlier Western estimates that only 3% to 5% of all goods exported to the East are subject to export controls, it will tell us only what percentage of actually realized exports is directly affected by the existing system (delays in deliveries, restrictions of re-exports, or exports of goods produced on imported technology, etc.). To that we should add trade that is not realized because of rejection of importers' applications for "exceptions" (in 1986-1987 there were around 25 rejections concerning export to Poland, most of them by the US). On top of all these, we should add an incalculable amount of trade not realized (and never registered) due to the fact that East European importers and/or Western exporters—conscious of existing restrictions—do not even apply for licenses.

14. Committee on Science, Engineering, and Public Policy of the National Academy of Sciences, the National Academy of Engineering, and the Institute of Medicine, *Balancing the National Interest: US National Security Export Controls and Global Economic Competition* (Washington, DC: National Academy Press, 1987).

15. To mention just two of them, one can cite examples of pipe laying and operating equipment and technology for an oil pipeline from the Soviet Union to Eastern Europe in 1963 and for a gas pipeline from Western Siberia to Western Europe in 1982.

strategically oriented longer-term export controls. Therefore, it is better to seek more complex analyses of this issue.

On one extreme, one can cite the results of a classic but controversial analysis by Gunnar Adler-Karlsson, who argued that a strict embargo imposed for a few years on exports to the Soviet Union caused practically no damage to the industrial and military potential of that country.[16] If that argument is correct, it would be even more unjustified to expect any real effects to that end later on, when restrictions became far less rigid. The authors of the most comprehensive analysis so far of international trade restrictions, (prepared in the early 1980s at the Institute for International Economics) reached similar conclusions concerning the effectiveness, or rather ineffectiveness, of COCOM controls.[17]

On the other hand, there is a relatively widespread view among Western experts that COCOM—though not effective enough in significantly slowing down the growth of the Soviet military-industrial complex—has been fairly effective overall in keeping the West's technological lead over the East.[18] And though that conviction is sometimes also questioned,[19] recent downward revaluations of Soviet industrial and military strength seem to confirm that broadly accepted view.

On the basis of the considerations mentioned here, one can draw the following conclusions: First, Western export controls have had only a limited effect on restricting East European access to the newest

---

16. Gunnar Adler-Karlsson, *International Economic Warfare 1947- 1967: A Case Study in Foreign Economic Policy* (Stockholm: Almquist and Wiksell, 1968).

17. Gary C. Hufbauer and Jeffrey J. Schott, *Economic Sanctions Reconsidered: History and Current Policy* (Washington, DC: Institute of International Economics, 1985).

18. See, among others, Ronald Amann, "Soviet Technological Performance," *Survey* 23, no. 2 (Spring 1978); Committee on Science, Engineering, and Public Policy, *Balancing the National Interest*; and National Research Council, *Global Trends in Computer Technology and Their Impact on Export Controls* (Washington, DC: National Academy Press, 1988).

19. For example, in *United States Military Posture for FY 1987*, Joint Chiefs of Staff Report (Washington, DC, 1987).

technologies of purely strategic (military) significance; however, these controls have significantly contributed to maintaining the overall technological lead of the West over the East. Second, the negative effects of Western multilateral export controls on proscribed countries have been concentrated mainly in the nonmilitary sectors of their national economies; this is especially true in the case of the Soviet Union, which has always tried very hard to shift its best resources from the civil to the military sector of the economy in order to keep up with the West in military strength. Third, Western restrictions have contributed to worsening the East European countries' economic problems and restricting their participation in the international division of labor, though it is practically impossible to calculate exactly the importance of export restrictions versus the importance of domestic constraints.[20]

The System in Transition

In the early 1980s, under the predominant influence of the United States, changes were introduced in the multilateral system of export controls. These changes, particularly the enlargement of the commodity scope of controls, raised many objections among COCOM countries.

Many countries suspected that the United States, by introducing these changes, was trying primarily to further its own strategic-political goals, i.e., the economic weakening of the Soviet Union and other East European countries, and not the officially declared security goals of the Western alliance. In addition, those restrictions seriously threatened the economic interests of many developed countries, especially in Western Europe.[21] Thus, it is quite understandable that Germany, France, Italy, Spain, Greece, and others often strongly protested

---

20. In chapter 3 of this volume, Volkhart Vincentz points out that those countries were prevented (through Western controls) from participating in intra-industry trade— the most dynamic sector of international trade over the last 20 years.

21. This was clearly shown, for example, in the case of deliveries for building the gas pipeline from Western Siberia to Western Europe. Especially harmful and controversial decisions were connected with the extraterritorial application and retroactivity of US laws.

against such changes and demanded essential reforms of the Western system of export controls.

These countries have frequently expressed their eagerness to continue mutually profitable trade and economic cooperation with Central and Eastern Europe. Their increasing interest in those markets has stemmed also from growing competition on the international market as a result of, among other factors, the steadily rising competitive position of exporters from the newly industrialized countries (NICs), mainly from Asia, but to some extent also from South America, and later on even from Africa.[22]

In the late 1980s, more and more arguments against an overly restrictive export control policy could also be heard in the United States. It has often been argued that such a policy creates an important barrier to further technological development in Western countries themselves, including the United States, and that the gains expected from the national security standpoint might not balance the losses suffered in Western entrepreneurship and research cooperation among Western countries.[23]

The overburden of export controls for the US economy has become more and more obvious as the US has lost, over the last 15 to 20 years, its position as undisputed leader of the Western world in many fields of technology. The level of technology has increased enormously not only in Western Europe and Japan, but also in many NICs. This rapid diffusion of technology is a result of a longer-term trend toward the internationalization of industrial processes and the globalization of international economic relations, as well as the increasing significance of direct international relations among enterprises, especially relations taking place within multinational corporations.[24]

---

22. According to the prevailing judgments, one should include in this category not only Hong Kong, Singapore, South Korea, and Taiwan, but also Malaysia, India, Brazil, Nigeria, and possibly other rapidly industrializing countries.

23. Committee on Science, Engineering, and Public Policy, *Balancing the National Interest*.

24. See, among others, Charles H. Ferguson, "America's High-Tech Decline," *Foreign Policy*, no. 74 (Spring 1989); Otis Port, "Why the US is Losing its Lead," *Business Week* (special edition), June 15, 1990.

Therefore, the recently changing approach of the US to the export control policy has been also a result of its worsening overall international economic position, and especially the position of its industry (including high-technology in particular) in relation to other countries. According to recent reports, in 1989—for the first time since World War II—another country (Japan) invested more than the United States did in plant and equipment, and by the mid-1980s the US had become a net importer of high-technology products.[25] This also contributed to its huge trade deficits, which have recently accounted for well over $100 billion annually (the highest deficit amounted to $162 billion in 1987; it was still over $100 billion in 1990).

There is no doubt that the deepening economic and political integration of Western Europe, especially the 1992 program, has also played an important role in influencing US export control policy. Analysts expect a relatively quick achievement of an almost entirely integrated market (comparable in size to that of the United States) for commodities, capital, and labor flows. This also means that numerous additional research projects will be undertaken by the European Community and all West European countries.[26] Both trends, closely interconnected, should mean the further economic and technological independence of that group of countries from the United States. Under such new conditions, the weakest links of the chain will decide the effectiveness of Western export controls;[27] thus, it has become extremely important for the United States to keep abreast of technological developments in Western Europe and to work out with the other Western countries a joint, unified policy of export controls to the East, even at the cost of some concessions from its own side.

---

25. Council on Competitiveness, *Competitiveness Index* (Washington, DC, 1990).

26. Those programs include the European Program for High Technology Research and Development (EUREKA), European Strategic Program for Research and Development in Information Technologies (ESPRIT), Research and Development in Advanced Communications Technologies for Europe (RACE), Basic Research in Industrial Technology for Europe (BRITE), Biomolecular Engineering Program (BEP), and European Research in Advanced Materials (EURAM).

27. According to some Western experts, this means primarily Italy, Greece, Spain, and Portugal.

As mentioned above, in the late 1980s, more and more arguments appeared in the United States in favor of changing the US and multilateral export controls, this time in the direction of liberalization. This reform-minded attitude has been especially strengthened as a result of the political and economic processes that started in Central and Eastern Europe in 1989. Though not equally advanced and evenly successful in all countries, the genuine efforts towards democratization, market-oriented economic reforms, and openness to the West have created an entirely different situation in that part of Europe. The Warsaw Treaty Organization (WTO) has been transformed from a military bloc to a predominantly political institution. One cannot exclude the possibility that it will be completely dissolved in a few years' time. This means that the fundamental rationale for applying Western export controls to the East has been seriously weakened recently or (in the case of some Central and East European countries, namely the CSFR, Hungary, and Poland) even significantly altered, since these countries no longer pose any strategic threat to the West.[28] In order to support the political and economic changes taking place there, the United States and other Western nations should not only stop restricting but should actively support the flow of many technologies to these countries.

The recent changes in Central and Eastern Europe have supplemented the overall improvement in East-West relations that started a few years ago. The first stage was marked on December 8, 1987, with the signing of the first agreement ever between the United States and the Soviet Union on the reduction of existing medium-range nuclear missiles. The signing of that agreement, which opened a new era in East-West relations, and highly advanced talks on the radical limitation of strategic nuclear forces (START) created a favorable atmosphere that also influenced trade and economic relations between East and West, including the transfer of technology. We have already seen a few examples of this new attitude, e.g., COCOM's decision in 1988 to allow sales of the newest generation of Airbus and Boeing airplanes to

---

28. One can even argue that, while the level of threat from Eastern Europe has substantially diminished, new security threats for the West have emerged. These are certain developing countries (primarily, but not only, from the Middle East) that have produced or acquired nuclear, chemical, and/or biological weapons.

the GDR, Hungary, Poland, and Romania. Since that time, restrictions on exports of computers and some other items have also been gradually eased.

Under the influence of all these factors, the United States—for years the strictest member country of the system—has appeared as a real champion of certain liberalizing changes that have recently taken place in the system of Western export controls. Starting in January 1990 with a very thorough review of its own export control policies, through an inconclusive COCOM meeting the following month, the Bush administration finally arrived at a number of recommendations for modernizing COCOM, which were presented and broadly accepted at a high-level meeting of COCOM in June 1990.[29]

Today, Western export controls undoubtedly face a period of the most critical changes since their inception. Nevertheless, one can still question whether the initiatives undertaken so far are radical enough and whether US policies bear more of a defensive or an offensive character. There are strong arguments in support of the view that US concessions have come just in time to head off a collapse of Western cooperation in controlling exports of sensitive technology to Eastern Europe.[30] On the other hand, however, there is also the view that, through easing technology export restrictions, the United States wants to completely overhaul the system, eliminate its burden for the US economy, and—with the strong support of other Western countries—seize the unique opportunity to further the democratic political and

---

29. Examples of this new attitude have included: 1) as of July 1, 1990, liberalizing controls to practically all proscribed destinations to levels consistent with the People's Republic of China's status; 2) further liberalizing levels of controls and giving more favorable consideration to "exception requests" for countries representing a lesser strategic risk (i.e., the CSFR, Hungary, and Poland), provided they adopt COCOM-approved safeguards regimes; 3) liberalizing controls for three commodity sectors (computers, machine tools, and telecommunications); 4) deleting 30 items from the COCOM International Industrial List on July 1, 1990, and an additional eight items by August 15, 1990; and 5) developing a new "core list," to replace the International Industrial List. See Harrison and Holliday, *Export Controls*. This new core list was intended to be established at the end of 1990, but that deadline was not met.

30. See, for example, Glennon J. Harrison, "Export Control Reform and the Revolution in Europe," *CRS Review* 11, no. 5-6, 1990.

economic reforms in Central and Eastern Europe, as well as encourage a movement in the same direction by the Soviet Union.[31]

One can question how the changes introduced in June 1990 will influence the future evolution of Western export controls to Eastern Europe, including the functioning of the COCOM organization.[32] First of all, these changes will mean a substantial reduction in the number of commodities subject to controls, especially in the commodity groups with a relatively lower level of technological advancement. However, with the liberalization of a relatively large group of technology exports to Eastern Europe, we can expect the strengthening of the export control regime and organizational functioning of the system where the most highly advanced, sensitive products and technologies are concerned.[33] This is a "higher fences around fewer products" approach. The proposed "core list" is expected to include eight categories of technologies: telecommunications; computers; navigation and avionic systems; sensors, sensor systems, and lasers; electronic design, development, and production; advanced materials and material processing; and marine technology.

Furthermore, one can expect that COCOM will maintain its strategy of keeping a technological lead over the Soviet Union and at least some other East European countries. Nevertheless, in view of the various political and economic changes taking place in the proscribed countries, there will be greater diversification among country destinations. The outcome will be the continuance of the policy of restricting

---

31. This view is supported by congressional initiatives, especially in the House of Representatives, to radically change the US law on export controls. Nevertheless, Congress and the Bush administration are not yet fully in agreement on export control reforms, as the latter prefers to go more slowly in that direction and retain its prerogatives in formulating US export controls policy. See Glennon J. Harrison, *Export Administration Act Reauthorization: House, Senate, and Administration Activities*, CRS Report for Congress, July 6, 1990.

32. There is no doubt whatsoever that, after the deeply disappointing experiences of the 1970s, the simple repeat of a detente-type policy in technology transfer to Eastern Europe is not possible. It would also be unwise, however, to expect the complete dissolution of Western strategic export controls in the near future.

33. See Harrison and Holliday, *Export Controls*. Nevertheless, there are still doubts, especially within US industry, that the new list will be a genuinely new, zero-based effort, as it is said now, or just a rewriting of the existing US Commodity Control List.

access to the most modern technology for the most rigidly controlled countries, with all its earlier-mentioned economic consequences. It is worth stressing, however, that in many cases the most important technologies come not from the US but from Western Europe, including non-COCOM countries. These countries have always constituted the most vital sources of technology imports from the West for practically all countries of Central and Eastern Europe, as well as for the USSR.[34]

As a result of the changes made in 1990, the export control systems of all Western countries will be more unified. This will mean, first of all, that the scope of US national controls will be cut further and moved closer (if not yet fully) to the multilaterally accepted level of controls. At the same time, restrictions on technology flows among COCOM member countries will be practically abandoned.[35] On the other hand, sanctions against persons, enterprises, or even countries not obeying the accepted rules would be increased and more strictly imposed.

Altogether, we will have to deal with a system of export controls—both in COCOM as a central body and in the national control systems of individual member countries—that will be organizationally and financially strengthened. These controls should also be more effective in protecting the strategic interests of Western countries. And that should mean cooperation among a growing number of countries in the West and most probably in the East in the implementation of strategic export controls, especially in view of the newly emerging threats to Western security interests.[36]

---

34. A few months ago, one could even hear from US officials the opinion that the United States was less inclined to liberalize export controls than, for example, West European countries, because its participation in the consequent transfer of technology to Eastern Europe would be meaningless. This view, however, is open to question, and East European (especially Soviet) buyers remain confident in the predominance of US technology.

35. The adoption of a "common standard" among COCOM countries for licensing and enforcement should lead to the establishment of a COCOM-free zone.

36. It is worth stressing, however, that COCOM is completely unprepared to meet these new challenges. So, in fact a "new deal" is needed that could mean less East-West and more North-South oriented export controls. See, among others, Quigley and Long, "Export Controls."

In view of such systemic changes, the ability of the United States to use the COCOM system of export controls for furthering its own foreign policy goals should be further diminished. One can doubt, however, that the US will forever refrain from pursuing such goals, especially in periods of deteriorating overall East-West relations (since no one can be assured of a completely problem-free future).[37] Nonetheless, in the scenario presented above, COCOM should evolve more and more into an entirely strategically oriented organization in its export controls to Eastern Europe. It could probably have that character exclusively if it transformed itself into a formal international organization, but this does not seem likely just now.

The scenario presented above of the evolution of COCOM and the entire system of Western export controls is not only possible, but in fact is already being implemented. One can only add to this forecast that—in the event of further progress in disarmament talks and the continuing improvement of East-West political relations—competition among countries will take place more and more in the sphere of civilian, not military, technology.

A Longer-Term Approach— Six Recommendations

To propose the immediate and total scrapping of Western export controls and the dissolution of COCOM would be premature at this stage of East-West relations and world affairs. However, comprehensive efforts should be undertaken both at international forums and in bilateral negotiations between individual Eastern and Western countries to rewrite export control rules and their application in East-West trade. Those efforts may have a political, economic, and organizational character, they may be more general or very specific, and they may have both short- and longer-term perspectives. Some efforts have already started, but their implementation should be strengthened and speeded up, and they should be supplemented by new initiatives.

---

37. Ibid. It may also happen that the possible failure of COCOM to accommodate newly emerging security threats from certain developing countries could force the United States to use foreign policy export controls to deal with that menace.

At the very beginning, it is worth stressing that—according to the East European view—the continuance of broad Western export controls has been mainly a result of the Cold War syndrome, i.e., the specific understanding by Western countries of their strategic and political interests in relation to Eastern Europe. Because of that, the existence of export controls has been in principle independent of the East European countries, especially the smaller ones. Nevertheless, though these countries have never had much say in the existence of export controls, they have been able to indirectly influence the commodity scope and the applied regime of those controls, insofar as they have been able to influence the present and future shape of overall East-West relations. The worsening of relations has always meant a wider scope and stricter regime of controls, while improvement has usually meant gradual limiting of the extent of those restrictions and a liberalization of the regime. We face now a complete redefinition of East-West relations and radical systemic changes in Eastern Europe, which should entail profound reevaluation of the West's understanding of its strategic and political interests in that region.

Therefore, the first of my recommendations is the very obvious proposition that Central and East European countries should—to the maximum extent and at all possible forums—aim at furthering the principle of mutually fruitful cooperation between East and West. Practical adherence to that principle by all countries will create the proper conditions for the peaceful (versus military) application of all products and technologies in the mutual interest of all partners in international economic relations. It should be widely understood that restrictions applied in international trade, especially those introduced for noneconomic reasons, cause damage to all countries involved and to overall economic and technological progress in the world, and thus should be eliminated to the maximum possible extent.

Such reasoning leads us to the second recommendation, which calls for the introduction of significant changes in the international regulation of export controls, at least in order to essentially limit such controls in peacetime. In other words, new international rules of technology trade should be worked out. To implement that idea, it is necessary for all Central and East European countries to engage themselves to a larger extent in the activities of international economic organizations and to coordinate their efforts with potential allied nations. That can be undertaken at various international economic organizations, such as GATT, the UN Conference on Trade and Development (UNCTAD), or the Economic Commission for Europe (ECE), or at other international forums, such as the Conference on Security and Cooperation in Europe (CSCE), which partially addressed the problem during its 1990 economic conference in Bonn. According to a recent report of the ECE:

> It is not clear whether or not these [COCOM] controls, intended to protect the strategic defense interests of the West, have seriously constrained the economic development of the East: it was suggested that the more binding constraints might be domestic in origin and especially those which reduced or removed the incentives for Eastern enterprises to innovate. However, if the various programs of reform are successful in weakening such systemic constraints, the bite of existing COCOM controls could become sharper and affect the development of strategic sectors, such as telecommunications. . . . It is clear that they constitute a potentially serious obstacle not only to the restructuring of industry and the emergence of new service industries, such as tourism, but also to the development of other activities, such as financial services and the dissemination of accurate and up-to-date economic information, which . . . are essential for the efficient working of a decentralized market system. Since the Western countries now perceive that they also have strategic interest in the success of economic and political reform in Eastern Europe, plans to relax a proportion of COCOM controls are already under discussion. . . . [They should introduce] a closer monitoring of the effects of COCOM controls on the Eastern economies and a more rapid review process whenever constraints on the Eastern reforms process are identified.[38]

Regarding economic assistance programs for the Central and East European countries, it has been stressed recently in the West that it

---

38. Economic Commission for Europe, *Economic Reform in the East*.

may be more important to provide them with the know-how and experience of running market economies than to give them only straight financial or other support. This leads to my third recommendation: the Central and East European countries must be allowed greater access to the most effective channels and mechanisms of technology transfer.[39] Usually included in this category are the purchase of turnkey plants, licensing agreements that provide personnel training for importers, joint ventures, agreements on the exchange of technology, training programs and university studies in foreign countries, access to foreign literature and foreign patents, etc. In the early 1980s, the Western countries, led by the United States, tried to limit the access of specialists from Central and Eastern Europe to new technology through some of those channels (by, among other things, not allowing them to participate in certain international scientific conferences). Today there is a need to change that policy and maybe even reverse it, i.e., to organize special conferences and training programs for various specialists from those countries. It has been recently suggested that there may even be a role in this for the Organization for Economic Cooperation and Development (OECD).

The fourth recommendation elaborates on one of the above, but it deserves separate attention—the necessity for greater involvement by foreign direct investment in Central and East European economies, including joint ventures. Generally, all those countries have had similar expectations from foreign direct investments, including hope for an inflow of modern Western technology and management techniques. However, more favorable conditions must be created in Eastern Europe for foreign capital to come and successfully function there. Such conditions include better regulations (provisions for unrestricted transfer of profits, easier access to host countries' resources, etc.) and the necessary infrastructure (telecommunications, transport, etc.). This can be done with some assistance from foreign capital. It should be firmly stressed, however, that infrastructure development cannot be left entirely to private capital to do the job. It should receive, at least at

---

39. See Philip Hanson, "The Import of Western Technology" in *The Soviet Union Since the Fall of Khrushchev*, ed. Archie Brown and Michael Kaser (London: Macmillan, 1976).

the beginning, some encouragement and support from Western governments.[40]

The fifth recommendation can be formulated as the necessity for Central and East European countries to open their economies and enter much more actively into the international division of labor, especially international technology trade. This involvement would create much more symmetric and stable economic relations between East and West. Asymmetry has always been one of the main characteristics of East-West economic relations: the Central and Eastern European countries are much more economically dependent on the West than vice versa, especially in the field of technologically advanced products. That asymmetry has always created a potential and quite often real risk of endangering the economic security of the Central and East European countries. To achieve greater symmetry, it is necessary for the East to first of all increase exports to the West, especially of industrial products. In fact, this is one of the most important elements of the far-reaching economic reforms undertaken in some Central and East European countries (primarily the CSFR, Hungary, and Poland). However, in order for these countries to succeed in that respect, the West must eliminate all barriers to trade. Even the eventual success of the Uruguay Round of the GATT negotiations, which is by no means certain, will not solve all the problems in question, so there will be still many problems left to be solved through bilateral arrangements.

In order to diminish the abovementioned asymmetry, especially in the sphere of technology trade, it is necessary for the Central and East European countries to develop their own bases of high technologies and to create and develop new export specializations in high-technology products. Of course, it would be unwise to expect that to happen very quickly. Nevertheless, they do not have to start from scratch, since those countries have their own significant technological achievements, which could be developed and properly marketed. To

---

40. The reasons are simple: the negative experience of the past, high indebtedness at least in some of those countries, and the inevitably long period of transition to political democracy and market economies, which entails some political chaos and instability. All these factors, and probably many others, make it difficult to expect Western private capital to enter those markets with huge investments.

do just that—and this is the sixth recommendation—they should apply for and be allowed to enter international research programs aimed at developing new technologies, especially projects being carried out in Western Europe. Some East European countries have already set a precedent in that respect.[41]

The abovementioned recommendations can help bring a satisfactory solution to the problem in the long term, but they are not enough to introduce the needed rapid and radical changes to the practice of export controls in East-West relations. To achieve that goal, additional measures must be taken. Certain countries have already started the process—the CSFR, Hungary, and Poland have started negotiations (or their preparatory phases) with the United States and some other developed countries on radical changes of their status within the system of Western export controls.

A Short-Term Approach—Solutions

Given the present political and economic situation of most Central and East European countries, combined with the overall improvement of East-West relations, there exists a favorable environment for easing and limiting the scope of the technology export restrictions still applied by Western countries.

In order to contribute to the necessary changes in the US and all-Western export control systems, Central and East European countries should conduct bilateral negotiations with the US on liberalizing both its national export controls and COCOM export controls. In the latter case, such discussions should be supported by talks with representatives of other COCOM member countries, especially those most interested in reducing the number of proscribed countries and limiting the commodities subject to export controls on the basis of their own economic interests. To this end, the CSFR, Hungary, and Poland started their negotiations with the United States in the spring of 1990, followed by talks with other COCOM countries.

---

41. A few Yugoslav, Hungarian, and Bulgarian enterprises joined some of the EUREKA projects in spite of a 1985 recommendation of West European countries not to include Eastern Europe in their work.

It is no wonder that, in the present favorable political conditions, US and all-Western technology exports to most Central and East European countries have been receiving more liberal treatment in COCOM since July 1, 1990.[42] At present, the goal of trade liberalization can be pursued by seeking more liberal treatment within COCOM, by negotiating agreements to sustain this treatment, and in the longer term, by working for fundamental changes in US export control law.

The CSFR, Hungary, and Poland—because of their internal democratization processes and economic reforms, as well as their growing political independence from the Soviet Union—are already receiving more liberal treatment than has been the case so far (and in comparison with the Soviet Union or some other proscribed countries). This has been achieved within the still existing legal and administrative systems of the US and COCOM export controls, through the US differentiation policy and through more liberal treatment of COCOM "exception requests" involving exports to these countries.

Gaining a more privileged position within the US and COCOM export controls system is, of course, worth working for. Nevertheless, one has to acknowledge that for Hungary and Poland a similar privileged position—though to a much lesser extent—had already been obtained (except during the late 1940s, early 1950s, and early 1980s, when the degree of differentiation was none or very small).[43] This privileged position, though it represents a step forward, is not satisfactory enough. It is too dependent on political changes that may

---

42. However, the scope of this liberalization, as prescribed in the June 1990 COCOM decision, was still not fully implemented as of the beginning of 1991. The ongoing differences between the US and the other COCOM member countries on the extent of liberalization, especially in the area of computers and telecommunications, contributed to the postponement of the high-level COCOM meeting planned for late February 1991.

43. While most of the proscribed European countries belong to the more restrictive country group Y (including the Soviet Union) within the US export control system, Romania (country group Q) and Hungary and Poland (country group W) have usually been treated more liberally (with the exception of Poland between 1982 and 1987). After the decision of May 27, 1989, all East European countries and the USSR received similar treatment. See *Export Administration Annual Report FY 1989* (Washington, DC: Department of Commerce, 1990).

take place in overall East-West relations, as well as in the United States itself, which still has a decisive say within the entire system. The policy and the practice of export controls in the United States, and to a large extent in COCOM, depend on the relations between different administrative departments (Commerce, State, Defense, and even Treasury, which supervises the Customs Service) and their actual importance within the system (and vis-à-vis the US Congress).[44] Any change of the US administration or in the internal division of power between the US Congress and the administration (and within that administration itself), or even simple wariness on the part of politicians because of a prolonged lack of full political and economic Western-type stability in Central and Eastern Europe in the foreseeable future, may significantly diminish the willingness of the United States to apply a more favorable regime of export controls toward that region.

It is necessary to strengthen and stabilize a more liberal status for those countries within the US and COCOM export control systems. It is possible to achieve that goal by negotiating bilateral agreements with the United States, in which importing countries would guarantee that they will obey the existing US rules on re-export of imported technology (mainly to the Soviet Union) and its utilization only for the envisaged purposes (preventing its military application). Such agreements should also provide for a mechanism of verification by the US authorities of importers' compliance with those rules. Generally, these agreements could be modelled on already existing strategic trade agreements negotiated by the United States during the 1980s with a group of non-COCOM countries (see below for details on a possible agreement).[45]

---

44. See William J. Long, *United States Export Control: Executive Autonomy vs. Congressional Reform* (New York: Columbia University Press, 1989).

45. These have been negotiated under the "third country initiative" (section 5 (k) of the Export Administration Act of 1979, as amended). However, the agreements reached so far with close to 15 countries differ significantly in form and content from one country to another. Some of them are close to introducing a COCOM-like regime of controls in non-COCOM countries, while others cover only a small part of COCOM controls. See Gary K. Bertsch and Steven Elliott-Gower, "American Perspectives on COCOM: Past, Present, and Future" (paper presented at the conference on "Forty Years After: COCOM, Alliance Security, and the Future of East-West Relations," Leiden (Netherlands), November 10-11, 1989).

Though the US still argues that such a solution is risky as long as those countries belong to the WTO or at least while Soviet military units reside on their territories, the situation is changing very quickly. As mentioned earlier, the WTO no longer functions as a military organization, and one can imagine that NATO will also evolve into a more political entity in the near future.[46] Furthermore, since July 1, 1990, we have already seen in East Germany the precedent for a country's exclusion from the majority of COCOM controls, even while Soviet soldiers still resided on its territory.[47] And with the unification of the two German states in early October 1990, there are Soviet soldiers now residing in a COCOM member country.

Nevertheless, even the signing of such agreements may not be enough for Central and East European countries or for US businesses to protect themselves against sudden changes in US export control policy in the future.

The US Congress is progressing with work on the extension of the Export Administration Act of 1979 and later amendments. The Act expired on September 30, 1990, but whether or not it is extended, the discussions on formulating an entirely new law will continue. This means that there is still time for US industry to lobby Congress and the administration for changes in the existing law's profile, especially its fundamental principle that exporting is only a privilege and not a fundamental right of US enterprises. The battle over the new shape of the export administration law is very important, as its outcome will most probably determine US exporting conditions of high technology for the coming decade. It is, of course, understandable that the direct engagement of foreign countries in the creation of a US internal law is not possible. Nevertheless, the extensive congressional debates, including deliberations in different committees of both the House of Representatives and the Senate, provide outsiders with numerous

---

46. At its July 1990 meeting in London, NATO leaders decided to move the organization away from its "forward defense" and "flexible response" strategies.

47. India is another example of a country with Soviet military advisers (and Soviet military equipment) being treated relatively liberally in the US and COCOM export control systems. There is even a joint venture on supercomputers established with the participation of Western capital.

opportunities to present their views (for example, at hearings with the participation of experts from different areas of industry, foreign countries, national groups living in the United States, etc.).

The steady progress in elimination of the "US differential" means that a multilaterally controlled list of items and the COCOM regulations will soon constitute the only body of the international export control system. Therefore, it is necessary to significantly reduce the number of proscribed countries and limit the commodities subject to COCOM controls. The leading position of the United States in that organization, assuming its genuine willingness to make essential changes in export restrictions, may significantly help in curtailing controls, but concentrating only on the US export control system is not enough. Central and East European countries must actively take advantage of the favorable attitude towards the reduction of controls on the part of at least some of the other COCOM member countries, which have never fully accepted the more restrictive philosophy of the US system of export controls. Many of them, mainly Western Europe and Japan, are genuinely interested in increasing their trade with the countries of Central and Eastern Europe, especially because the political and economic processes taking place in the latter have opened new opportunities for increased exports from the West, including export of technology indispensable for modernizing, restructuring, and reviving the East European economies.

It is obvious that the expected changes in the US and all-Western export controls will not happen overnight, but a timetable for their introduction may be worked out soon. Of course, much depends on economic and political developments in Central and Eastern Europe (as well as in the Soviet Union). The prospects for political change look relatively good, but there may be more difficulties with successful implementation of economic market-oriented reforms. One has to remember, however, that such reforms themselves should work in favor of relaxing Western export controls. The growing decentralization of economic decision making should make enterprises more and more responsible for all their activities, including any export or re-export decisions of imported technology. And they, in their own well-understood interest, will be more and more interested in protecting the rights of their Western suppliers and/or cooperation partners. Therefore, the market-oriented economic strategies of most Central and East European countries constitute an additional favorable factor for the creation of an effective safeguards system against unauthorized use of imported technology.

A Technology Safeguards System

Under an agreement signed with the US (and possibly accepted by other Western countries and/or COCOM) a Central or East European country may create a safeguards regime, the scope and restrictiveness of which would depend on the scale and level of control still applied to that country.

The technology safeguards regime would be directed at preventing the diversion of US-origin industrial list controlled strategic goods and technology and, further, of COCOM-origin industrial list products contrary to the security interests of COCOM members, whether such diversion were to take place within a proscribed country or by re-export from that country. The system of safeguards should be flexible enough to extend at certain moments comparable protection to controlled strategic goods and technology originating in COCOM-cooperating and even other third countries. The establishment and practical implementation of the safeguards system could take place in several steps (at least two), assuming that further steps toward a more stringent regime would result in further easing of export controls by the West. The introduction by any country of a full-fledged (Western standard) safeguards system should result in its final removal from the COCOM proscribed destinations list (or the dismantling of that organization) or at least an agreement establishing a concrete timetable for that removal.

The initial introduction of a safeguards regime could be generally based on a country's existing legislation (economic activity act, customs law, penal code, etc.), supplemented by administrative orders, and need not necessarily involve any parliamentary action.[48]

---

48. Though one can also imagine that a special law on controlling technology export could be enacted at this early stage, as was proposed by the CSFR in its negotiations with the United States.

Therefore, it could be introduced relatively quickly, i.e., within a few months. This stage could have all or most of the following characteristics:

- The introduction of import certificates and delivery verification procedures for imports of controlled goods and technologies. This would contain the importer's commitments: a) to import the goods directly to his or her country and within the quantities specified in the certificate; b) not to divert the goods to another destination through transshipment or re-export without prior consent of the supervisory entity (ministry of foreign trade or another office established especially for the purpose); c) to notify that entity of any substantive changes in the information presented in the certificate; d) to allow pre-license and post-shipment checks to be performed on the importer's premises by his or her country's customs officials, assisted by foreign (exporter's country) representatives. The import certificate should be submitted to the supervisory entity for certification; and the delivery verification form should be submitted by the importer to the appropriate customs office for certification that controlled commodities entered the appropriate customs area in accordance with the law.

- Sanctions for knowingly making false statements during the course of proceedings conducted under the law or for providing misleading information in order to obtain a document or certification should be set at an appropriate level, generally using already existing law (penal code); sanctions should be harsh enough (including prison terms) to prevent or at least seriously limit the likelihood of any illegal action.

- The country's authorities should establish that an exporter has a legal/administrative obligation to obtain from a supervisory entity a license for the exportation of goods on the COCOM list; an exporter applying for such a license should submit a copy of the original foreign export license—if it imposed any restrictions on re-export, an export license should be denied, unless a waiver from the original exporter's government was obtained or conditions in the original export license were met.

- Provisions that authorize control over transit and in-bond shipments should be also included (if they do not already exist) in the customs law of every country; all goods entering respective customs areas and moved from abroad to customs-free zones should be subject to customs supervision and control; thus any breach of an importer's commitments under an import certificate can be identified and punished.

- Each country's law (penal or treasury law) should provide sanctions (a fine, a jail sentence, or both) for attempting unauthorized exportation in violation of regulations on export licenses; the same sanctions should apply to any person who knowingly acquires, stores, conceals, helps to sell, or transports goods with regard to which an infringement of the customs law takes place; a violation of export control requirements should also be grounds for denial of a future export license.

- Special information networks should be organized to make importers aware of the system in question, and a special office should be established to deal with all the inquiries and requests for import certificates and delivery verification forms; there should also be special training courses for government administrators and enforcement personnel who deal on a day-to-day basis with those issues.

- The conclusion of a customs-to-customs agreement between an individual Central or East European country on one hand and the US and/or other COCOM member on the other would constitute an important additional factor strengthening the practical implementation of the safeguards.

- In order to achieve a higher level of cooperation and efficiency of the system, the cooperating parties should periodically review the experience gained under the technology safeguards program.

It would be rational to introduce all of the abovementioned measures in a relatively short period of time in order to create a fairly comprehensive, though not yet full-fledged, system of safeguards at this early stage.

The second and possibly final stage would result in a comprehensive safeguards system. This would involve the introduction of a series of more advanced measures, comparable with those adopted in other COCOM-cooperating countries. That stage would involve, among other things, enactment of a high-technology trade law whose purpose would be twofold: a) to codify and refine all the existing elements of the safeguards regime established under stage one, described above; and b) to broaden both the scope of transactions covered by the safeguards system and the enforcement powers of the customs service of an individual Central or East European country.

It is understood that the removal of any country from the list of proscribed destinations would require the extension of safeguards to strategic goods and technologies originating in countries other than COCOM member or COCOM-cooperating countries. Such an extension should also establish the legal basis for restricting transit through a Central or East European country and its customs-free zones of COCOM-controlled goods and technology acquired by third parties.

The proposed law should also provide that any passage of controlled goods and technology that would take place within a country between its own and foreign or other unauthorized persons would be subject to the safeguards regime and requirements established thereunder. In order to monitor such transactions and to enforce any restrictions imposed, customs services must be granted power to investigate cases, seize controlled goods, and detain individuals involved in actions that have as their object illegal transfer of strategic goods and technology, but that have not yet reached a stage where current customs law has been violated. The law would also have to provide appropriate sanctions for these offenses.

The date by which the final stage of such a system could be implemented depends to a great extent on a US/COCOM decision to remove an individual Central or East European country from the list of proscribed destinations. Its government, in order to introduce such far-reaching legislation on a safeguards regime to its parliament, would need to identify tangible benefits that would offset the commitments it would have to make. The removal from the list of proscribed destinations would, of course, constitute such a convincing benefit.

It is fully understandable that experience gained during the early stages of the safeguards system would certainly play a very important role in the decision-making process on both sides. Assuming that everything goes well, it would be feasible to introduce the final stage of the safeguards system by the end of 1991 (e.g., in Poland, if the decision of a binding commitment to remove it from the proscribed destinations list is known by that time).

There could be, however, other factors that could delay that process. One can name just a few: larger political tensions between East and West that could prolong uncertainty and increase Western reluctance to give up an important tool of economic leverage in relations with the East, prolonged difficulties with the successful transformation to market economies, or negative experiences in the early stages of the introduction of safeguards measures. The interested Central and East European countries hope, however, that this will not be the case, and that the process of their removal from the list of export controls proscribed destinations will proceed smoothly.

Conclusion

There are strong arguments in favor of a radical change in the system of Western export controls:

- The characteristics of Western technology export controls to Eastern Europe (the broad scope of controls, the system's rigidity, and the high economic costs for all countries involved, including Western economies), on one hand, and the changing perception of the security threat for the West from the East, on the other;

- The necessity for most of the Central and East European countries not only to continue, but to significantly increase their imports of technology in order to fully transform their economies into market-oriented ones, modernize most of their industries, and improve their products' competitiveness on the world market;

- The general perception that the far-reaching relaxation of export controls from the Western side may very significantly contribute to the successful outcome of the processes taking place in all post-socialist countries, and—on the other hand—that the successful implementation of economic and democratic reforms in Central and Eastern Europe could have a crucial impact on the Western policies of export controls in the future.

From these factors, two important conclusions may be drawn:

- First, it is necessary for all parties involved to undertake very active and comprehensive efforts to radically change the system of Western export controls and eliminate or essentially diminish their potential negative effects for economic and political developments in Central and Eastern Europe.
- Second, concrete possibilities now exist for the effective cooperation of Central and East European countries with the West in enforcing export controls on the transfer of technology to other destinations that pose real security threats to the West and to all democratic countries.

3

Western Export Controls, International Technology Markets, and the Performance of the East European Economies

VOLKHART VINCENTZ

COCOM, Ostpolitik, and the Prospects of Unification

When the wall crumbled in Berlin, German reunification loomed imminent and began to dominate politics in the Federal Republic of Germany. Relations with Eastern Europe and the Soviet Union gained importance, and the cooperative element that had long been part of the FRG's policy towards the East was strengthened. This policy was to a large part determined by Deutschlandpolitik, which concentrated mainly on humanitarian improvements for the citizens of the German Democratic Republic. Having started with Ostpolitik in the 1970s, the policy was basically cooperative in nature; it aimed to create a web of relationships and mutual responsibilities between East and West. The resulting stable and growing communication between the blocs was to

---

The author would like to thank the participants of the IEWSS colloquium for helpful comments.

eventually let the ideas of a democratic, open society take root in the East European societies. Trade was an important area where this policy was applied.[1] The slogan "change by means of trade" (*Wandel durch Handel*), used in the 1970s in FRG politics, stressed the special importance attached to commercial links. Trade incentives, rather than sanctions, were favored in pursuit of the FRG's objectives in Eastern Europe and the Soviet Union.

The major political parties in the FRG interpreted the changes in Eastern Europe, especially the events in the GDR, as a success of Ostpolitik.[2] Thus, the continuation of the cooperative approach in dealing with Eastern Europe was emphasized. Public opinion in the FRG at the end of 1989 and in the beginning of 1990 was also largely in favor of cooperation with Eastern Europe. For most politicians, the logical consequence was to nurture further the ties between East and West. In this environment, easing the impediments to trade caused by the COCOM regulations was on the agenda.

Proponents of Ostpolitik believed that the German question could not be solved without Moscow. After November 1989, it seemed possible that the Soviet leadership would agree to German unification. To reduce the chances that this opportunity would be missed because of unpredictable changes in the Soviet leadership and politics, it was necessary to accelerate the process of unification to the utmost, and to achieve as quickly as possible the consent of the Soviet Union. Offers of generous credit facilities and relaxation of trade regulations were brought in to smooth the negotiations. Given the great importance the Soviets attached to the question of COCOM, West German lobbying to ease curbs on technology must have been favorable for the talks between Moscow and Bonn on unification.

The longstanding question about the effects of COCOM regulations on the economy was, in the case of the GDR, fast and

---

1. The West German view of COCOM and East-West trade is described in Hanns-Dieter Jacobsen, "East-West Trade and Export Controls: The West German Perspective," in *Controlling East-West Trade and Technology Transfer: Power, Politics and Policies*, ed. Gary K. Bertsch (Durham and London: Duke University Press, 1988), pp. 159-92.

2. Of course, politicians who followed a policy of strength and confrontation against Eastern Europe also claim that it was their policy that ultimately brought down totalitarianism in the East.

convincingly answered. When the first Western entrepreneurs visited GDR firms to discuss cooperation arrangements, they were quick to point out that only a major inflow of Western technology could prevent the collapse of the East German economy. The politicians followed suit in stressing the necessity of modern technology for the transformation of the GDR into a market economy. West German public opinion on East-West issues, the indispensable support of Gorbachev for German unification, and the necessity of modern dual-use technologies for the reforming East European countries, especially the GDR, explain why the government of the FRG favored a rather substantial liberalization in the June 1990 negotiations on revisions of the COCOM list.[3]

As a result, COCOM restrictions were almost completely abandoned for the GDR, starting with the establishment of the monetary union between East and West Germany on July 1, 1990. Only a small group of military goods, the so-called bikini list, was still banned for export to East Germany. Even though the GDR did not become a member of COCOM, its government gave assurances that neither machines nor know-how would be passed to third countries. A control system was established to police technology trade. The transfer of technology originating in East German firms, however, was not restricted, so that the GDR could fulfill its export obligations to the USSR, which the FRG has pledged to honor.[4]

The slogan for adjusting the COCOM list, which was agreed upon by most of the parties, called for "higher walls around fewer items." By June 1990, however, the complications in monitoring the COCOM regulations must have been obvious. Considering the close economic entanglement of the Soviet Union with Eastern Europe in the past and the still existing and functioning personal contacts, it is doubtful whether the new COCOM regulations really can establish "higher walls" around the reduced list. It was for reasons of insufficient control that the United Kingdom had resisted, at the beginning of the negotiations, any relaxation of the technology controls at all. Finally,

---

3. In December 1989, the Social Democrats, the opposition party in the West German parliament, had demanded the abolition of all controls for dual-use technologies.

4. With the unification of the two German states in October 1990, all export restrictions to eastern Germany were abandoned. COCOM regulations on exports from the unified Germany are enforced as they had been in the FRG.

however, the COCOM members reduced the list of items to be controlled, and seemed to have tacitly accepted a weaker surveillance of the regulations, which naturally results from more open borders between East and West. Trade restrictions on three items—machine tools, computers, and telecommunication equipment—were greatly liberalized. These products are seen as the most crucial for modernizing East European economies. Thus, the revision of the COCOM list in June 1990 can be classified as a major liberalization that should lead to an increased technology flow from West to East.

In addition to the GDR, preferential treatment was given to the CSFR, Hungary, and Poland. In order to obtain this, these countries must set up control systems of their own to safeguard sales of strategic goods and technology. In addition, Western observers will carry out on-site inspections in these countries.[5] Restrictions were also significantly eased for the other countries, including the Soviet Union.

These recent changes made in order to facilitate East-West technology transfer may be only the beginning of an almost complete repeal of East-West export controls for dual-use technologies, if Eastern and Western Europe improve their political, cultural, and economic ties and move towards an integrated Europe. But even if this occurs, it is useful to look at the hindrances to technology transfer in the past to learn about the preconditions for an efficient technology transfer to the reforming economies in the future. Furthermore, the issue of controlling technology transfer will remain on the agenda even if the East-West confrontation dissolves. The experiences of the past might be useful in assessing curbs on technology flows into other parts of the world.

Effects of COCOM on the Civilian Economies of Eastern Europe

Introduction

Judging from the rhetoric used by Soviet officials to attack the COCOM list, its negative impact on the efficiency of the Soviet

---

5. *Financial Times*, March 6, 1990.

economy seems rather substantial.[6] A similar impression is given if one listens to complaints from Western industrialists. On the other hand, many Western analysts believe that the Soviets exaggerate the effects of COCOM on their economic performance, and that the commercial community unduly deplores the missed business opportunities. For several reasons, then, it is time for a fresh look at the harm the export restrictions did inflict and could possibly inflict on the economies of Eastern Europe.

Although most discussants dismiss the idea that economic warfare is an objective of COCOM, there was not much motivation for concern about COCOM's impact on civilian economies during the period of high tension between East and West. However, the reforms in Eastern Europe and the USSR have completely changed the perspective on this issue. It is no longer in the interest of the West to destabilize or impede reforming economies such as Poland, Hungary, the CSFR, or the GDR. A successful transition from a planned to a market economy would be of advantage to the West as well as to those countries themselves. Because rather large amounts of Western money are being poured into the reforming economies, donors must take care that the desired effects of credits and aid are not diminished by restrictions on technology transfer.

In addition, sweeping economic reforms in socialist countries demand that several essential measures be implemented simultaneously and quickly. The monetary union between the FRG and the GDR demonstrated this convincingly. It was strongly recommended, on both political and economic grounds, that the GDR's economic system be transformed as quickly as possible. The same holds true for the other reforming countries of Eastern Europe. The West must make sure that the development towards democracy and market economy becomes irreversible. All this adds a crucial time element to the value of any support from the West. Once reforms have started, the need for certain technologies arises, which, if not satisfied, may easily be a factor that delays the whole reform process.

Finally, the growing importance of technology as an engine of

---

6. Most notable is former Foreign Minister Eduard Shevardnadze's reference to the "damned COCOM list."

Table 1
Percentage of High-Technology Exports in Exports of Manufactured Goods from Industrialized West

Recipients	1970	1980	1981	1982	1983	1984	1985	1986
USSR	18.1	15.2	12.5	13.7	14.6	12.4	12.7	12.6
Eastern Europe	14.9	15.3	15.4	16.8	16.6	14.5	15.1	16.9
China	9.8	12.0	12.9	11.8	11.7	12.5	14.8	20.1
World	14.9	15.1	16.2	17.5	18.3	18.7	18.7	18.8

Source: US Department of Commerce from UN Series D data.

growth calls for a new assessment of restrictions on technology transfer. The significance of technology for the prosperity of a country has already stimulated a broader definition of national security. According to B.R. Inman and Daniel F. Burton, Jr. of the Council on Competitiveness, "National security can no longer be viewed in exclusively military terms; economic security and industrial competitiveness are also vital considerations. Indeed, where technology is concerned it is difficult to tell where military concerns stop and economic issues begin."[7] If this is true for the US, it also has significance for the other superpower, the USSR: participation in the international exchange of technology is as vital for the Soviet Union as it is for all other countries.

East-West Technology Trade

This section examines the extent to which Eastern Europe has participated in international technology trade, giving clues to the potential losses these countries experienced from their separation from the world market. Table 1 shows the share of high-technology exports in total Western manufactured exports to different regions.

The share of high-technology exports to the USSR declined substantially during the 1980s; in the case of Eastern Europe, the share remained almost constant. In contrast, the high-technology share in

---

7. B.R. Inman and Daniel F. Burton, Jr., "Technology and Competitiveness: The New Policy Frontier," *Foreign Affairs* 69, no. 2 (Spring 1990), p. 133.

world exports increased significantly. This resulted mainly from the increasing exchange of high technology among the Western industrialized countries, which reflects the acceleration of technological development and the expansion of markets for high technology.

From these data we can derive two conclusions. First, the share of high-technology exports to the countries of the Council for Mutual Economic Assistance (CMEA) increased significantly less than the exchange of high technology within the West. Second, in the mid-1980s the share of high technology in total exports to the CMEA was less than the share in the trade among Western industrialized countries. From a purely economic point of view, the latter result is surprising, because one would expect a rather high flow of technology from the technology-abundant West to the technology-poor East.

This outcome can be attributed largely to Western export controls. But other factors also affected East-West technology trade. The low share is partly a result of low East European demand for certain high-technology products. In particular, there was only a small demand for consumer electronics, which constituted a significant part of Western technology trade.

But despite demand-related factors, it seems that East European imports of technology were significantly restricted by COCOM. This can be seen from the composition of exported goods from the West. It is telling that computers and telecommunication equipment, for which there was clearly a desperate need, constitute only a small fraction of Western technology exports to Eastern Europe.

To get an idea of the obstructions to Western high-technology exports posed by COCOM, it is illuminating to compare deliveries to the People's Republic of China with deliveries to the CMEA countries. After export controls for China were eased, the share of high-technology exports to that country quickly increased, and by 1986 was well above the world average. This can also be seen if one considers the US share in total Chinese high-technology imports. In the early 1980s, about 12% of high-technology exports to China came from the US. Starting around 1983, the US unilaterally relaxed the controls for China by using exception procedures while the control list remained unchanged. The US share in high-technology exports to China increased from 21.2 % in 1982 to 30.6% in 1983. In 1984, exception requests in the US reached more than 3,000, which was 50% above the level at the beginning of the 1980s.[8] In 1985, COCOM rules for China were multilaterally eased, and, as Table 1 shows, the share of high technology increased, while the US share fell to 25%. This experience indicates strongly that a relaxation of the COCOM list will be followed by increased high-technology imports.

The example of China also demonstrates that an uncoordinated, unilateral change in the handling of COCOM restrictions, as was carried out in 1983 and 1984, can very well be used by a single state to gain an advantage over its competitors. The possible misuse of COCOM to circumvent fair competition among Western firms is a fiercely debated point.

A broader international comparison of high-technology trade gives further insight into the actual and potential technology trade with the East. Table 2 provides the necessary data.

Although the data in Table 2 do not account for intra-CMEA trade, it is clear that the East European countries are significantly less involved in international high-technology trade than all other trading partners. Only 7%-8% of their imports and exports are in high-technology products. The loss of competitiveness has resulted in rapidly declining export shares of high and advanced technology. This indicates that the technological gap between East and West increased in the 1980s.

The import data illustrate the close linkage between exports and imports of technology. An integration into the international division of labor demands substantial imports of technology. A higher inflow of Western technology seems to be a prerequisite for Eastern Europe, if its reformed economies attempt to capitalize on the gains of an outward-oriented development strategy.

Necessity for and Usage of Technology in the Socialist Economies

The data on East-West trade demonstrate clearly that in the 1980s the already low competitiveness of all socialist economies deteriorated

---

8. Michael Mastanduno, "The Management of Alliance Export Control Policy: American Leadership and the Politics of COCOM," in *Controlling East-West Trade*, ed. Bertsch, pp. 241-79.

Table 2
Share of High and Advanced Technology in Engineering Goods Trade*

	High Technology			Advanced Technology		
	1970	1980	1987	1970	1980	1987
	Exported Shares					
East European countries	8.8	8.4	7.8	23.2	19.7	16.6
4 Asian NICs	37.8	35.8	41.4	13.0	8.7	9.4
16 LDCs	26.3	28.2	57.2	10.6	15.9	10.6
Western Europe	14.7	16.1	18.4	15.3	16.1	15.1
United States	31.5	32.8	43.8	15.6	17.8	17.0
	Imported Shares					
East European countries	7.3	6.0	7.1	23.9	21.1	23.7
4 Asian NICs	19.0	33.1	39.4	17.4	17.0	17.9
16 LDCs	13.2	14.2	22.2	16.2	16.5	17.8
Western Europe	19.7	22.3	25.0	16.0	15.1	14.9
United States	14.0	22.2	24.7	11.4	11.5	11.8

*Production is classified as high-technology if the R&D expenditure exceeds 7.1% of total sales. In the case of advanced technology, R&D expenditure is more than 2.6% but less than 7.1% of total sales. US data were used to determine the high and advanced technology products.

Source: ECE, *Economic Survey of Europe in 1989-1990* (New York, 1990), table 7.4.1, p. 343.

further and the technological gap between East and West increased. The reasons for these developments are to a large extent inherent in the socialist economies. Although I acknowledge the importance of domestic factors for the poor state of technology in Eastern Europe, I will argue that export controls also inflicted damage on the civilian economy.

The experience of the past shows that the socialist economies absorbed modern Western technology only to a very limited extent. The new, imported machines and instruments were not used efficiently; therefore, it was argued, the nonavailability of high technology did not really matter, and the persistent call for high-technology products from many East European politicians has been related more to prestige than to economic needs.[9]

One explanation of the unsatisfactory absorption of technology can be found in the economic mechanism. The incentive structure in state-owned enterprises is biased against inventions and new production methods. The producer market and excess demand guarantee that almost everything can be sold. Thus, there was little incentive for product innovations. In addition, the policy of full employment effectively impeded rationalization. All this was responsible for the slow diffusion of technology in the socialist economies.[10]

As the East European countries develop further into market economies, however, the hindrances resulting from the economic mechanism will be reduced or will even vanish entirely. Whatever the domestic factors that contributed to the miserable state of technology, there remain also the restrictions from COCOM to be considered. It is possible that these restrictions did not really hinder the development of the civilian economies in the past because they were masked by the difficulties inherent in those economies. But even if that were so, once the market does away with organizational slack and misleading incentives, the potential effects of COCOM would become effective impediments to the development of the civilian economy.

There are many examples in Eastern Europe and the USSR showing that technologies, though available, are not used widely enough. An often-quoted example is the diffusion of oxygen converter and electric furnace processes in steel production. While in the West all steel is produced with these modern technologies, half the steel in the USSR is still produced with the outdated open hearth process. But it would be wrong to argue that in socialist countries the diffusion of all technology lagged far behind the West. The institutional reasons mentioned above were not the only factors that determined the speed of diffusion. Technology-specific factors also seem to have had a strong impact. There are examples, such as diesel versus steam

---

9. See R. Rohde, *Wieviel Exportkontrolle? COCOM auf dem Prüfstand* [How many export restrictions? COCOM on trial], Hessische Stiftung Friedens- und Konfliktforschung, Report 6 (1988), p. 24.

10. See Igor E. Artemiev's contribution in this volume for more details.

locomotives, where the new technologies proliferated rather quickly through the Soviet economy. One can also imagine that in a socialist economy personal computers, for example, would be in widespread use and would be of some value if a sufficient number of them were available. Their productivity would certainly be lower, however, because some synergy effects—from networks, for example—could not be achieved because of the miserable communications systems. Often the necessary infrastructure, the technological environment for the efficient use of high technology, is missing, and therefore the productivity of those technologies is low.[11]

Certainly, a fair number of high-technology products are of no use or very limited use in socialist economies. A computer network to enable a just-in-time system of immediate delivery is superfluous, if firms treat inventories as a medium of exchange to barter for needed goods. But there are other products and uses that would have a positive effect in a socialist economy: examples include a functioning telephone system, or CAD (computer-aided design) and CAM (computer-aided manufacturing) applications.

A general verdict that high technology is of no use in socialist economies would clearly underestimate the damage done to those economies by the nonavailability of certain high technologies. This is especially true for basic technologies such as electronic components, which enter into a host of other products. Without these technologies, e.g., modern chips to control machine tools, the chances of producing competitive machinery products for export are greatly diminished.

In the past, export controls had a very distinct effect on the civilian economy. Deprived of the opportunity to buy high technology on the world market, Eastern Europe was forced to produce it domestically. Great amounts of resources were earmarked for these purposes, but the effect was rather meager. The decision of the GDR to develop its own electronics industry is a good example. The GDR's role as a technological leader within the CMEA was achieved by huge investments that never made profits. The losses in this industry were partly the result of the high cost for the procurement of embargoed

---

11. It should be mentioned that the West has its own productivity paradox: "We see computers everywhere but in the productivity statistic."

chip-making equipment, for which up to three times the world market price had to be paid. But in addition, the production runs were much too short, and the waste too high to capitalize on declining average costs in production. A recalculation of the profits for different branches in the GDR showed that the electric and electronics industries were the least profitable. For every DM 100 revenue, this industry suffered 50 DM losses, if market prices had been used.[12] For the same reasons, the development of the much-praised megabit chip was stopped in the summer of 1990. The decision to develop a domestic semiconductor industry was the country's biggest misdirected investment.[13] At the moment, all signs indicate that those parts of the GDR electronics industry that were the most modern by CMEA standards will not survive in the open market.

Export controls also deprived Eastern Europe of some specialization opportunities that seem to have been well suited to the factor endowment of these countries. Some developing countries have to a large extent exploited their advantage of abundant labor, by assembling high-technology products, such as personal computers, or by developing software via telecommunication on computers based in Western industrialized countries. Although the processes themselves are not high-technology, the means of production that were utilized do involve restricted technologies, and therefore such assembling or software development could not be done in Eastern Europe or the USSR. The potential of such activities for Eastern Europe can be gauged using the figures on LDCs in Table 2. The strikingly high export shares of the less developed countries in the area of high technology presumably originate from such production, namely the assembling of electronic parts.[14]

---

12. *Der Spiegel*, May 21, 1990.

13. According to R. Hillig, the deputy head of the GDR firm VEB Mikroelektronik, the development of the megabit chip "was a purely political decision, and it turned out to be our ruin." Altogether, the GDR spent 30 billion East Marks on R&D for the semiconductor industry, half of it for the production of the 1-megabit chip in the last decade. *Financial Times*, June 27, 1990.

14. Economic Commission for Europe, *Economic Survey of Europe in 1989-1990* (New York, 1990), pp. 7-46.

The increase in high-technology exports to China after the relaxation of the COCOM list, as mentioned earlier, would also indicate that the use of technology in the civilian economy was restricted by export controls. In the Chinese case at least, the demand from the civilian sector for high technology certainly did exist.

Demand for and Financing of Technology in the Reforming Economies

In view of the rapid changes in Eastern Europe at the end of the 1980s, one must pay special attention to the fact that the reform process itself creates a demand for technologies.

A striking example of how reform in Eastern Europe increases the demand for high technology can be seen in the emerging money and capital markets. These markets must replace a significant part of the former planning bureau, which had decided on the allocation of investments. To do so, an efficient banking system is needed at the beginning of the reform. Not only must banks allocate capital, but they will also play an important role in evaluating the existing capital stock. This evaluation is necessary both for privatization and correct cost accounting in firms. Fulfilling these tasks requires much more efficient communication channels than those available in Eastern Europe. A banking system needs a large amount of computing power and modern telecommunications devices. In view of the urgency to create functioning money and capital markets, acquiring appropriate communications technologies will become a top priority. Generalizing from the example of the banking sector, one might argue that a decentralized market system can function only with efficient communications systems, the technology for which must be imported.

The demand for modern technology in the areas of environmental protection and safety will also increase. In the emerging democratic and open societies of Eastern Europe, the issues of pollution and safety can no longer be suppressed. Because of its geographical proximity, Western Europe must be seriously concerned about pollution and the low safety levels of East European and Soviet nuclear power plants. The pollution damage now evident in East Germany is much worse than estimated before; if FRG safety standards are used, all nuclear power plants in the former GDR will shut down.

The reforms will also change the financing of technology. Some experts argue that the abolition of export controls will not be followed by any substantial increase in East-West trade, because of the hard currency shortage, which will not disappear with the relaxation of the COCOM list. The best that can be expected, according to these experts, will be a substitution of certain imports by high-technology imports, although this substitution is restricted insofar as the existing imports fulfill basic needs (such as grain in the case of the USSR) or are in other ways essential to keep the economy running.

But in a way, technology imports might generate their own financing. All countries of Eastern Europe and the USSR are trying to attract direct foreign investment, and joint ventures should serve as an important means of proliferating modern technology in the domestic economy. It is true that the impact of joint ventures has been largely overestimated. At this time their contribution to the East European economies is rather small. If the reforms continue, however, conditions for joint ventures will improve. Technology transfer through foreign direct investment might live up to expectations, especially if the repatriation of profits is allowed.

In the same vein, the availability of commercial credits is linked to access to certain high technologies. If credits are placed on a project-oriented basis, instead of being linked to country risks only, the technology involved in a project begins to gain more importance.

Some of the COCOM-restricted technologies are typically used for infrastructure investments such as telecommunications equipment. In East and West, however, infrastructure belongs mainly to the domain of government activities. Therefore, there is a good chance that Western aid to East European governments will target such infrastructure investment. The World Bank, for example, has already given Poland a credit of $120 million-$150 million for the development of its telecommunications system.

East-West Technology Transfer and the International Division of Labor

This section looks from a broader perspective at barriers to East-West technology transfer. Several factors, in addition to Western export controls, conspired to severely limit the participation of Eastern

Europe and the USSR in the worldwide division of labor. Identifying the costs and causes of the weak integration into the world economy may help to assess the chances and risks faced by the East European economies in the future, when most curbs on technology trade could be lifted.

The East European economies were badly prepared for the rapid technological development the world experienced in the 1980s. This was demonstrated by the data in Tables 1 and 2, which show that Eastern Europe's participation in global high-technology trade was minimal. The lack of competitiveness in this area is evident. There are still other worldwide trends of trade that Eastern Europe could not follow.

The bulk of trade and the most dynamic component of exports among Western countries in the last decades has been the so-called intra-industry trade. Trade volume in the West has been dominated not by the exchange of raw materials for machinery (inter-industry trade), but by exports and imports of similar products (intra-industry trade) with rather high-technology intensities. While East-West inter-industry trade was governed by international differences in relative factor endowments and natural resources, intra-industry trade among Western countries exploited economies of scale in the production of specific items. According to one estimate, in the mid-1980s the share of intra-industry trade in total trade among the industrialized Western countries was over 60%.[15] Another study calculated that almost all (98%) trade among EC members was of this kind. On the other hand, only about 40% of the EC-CMEA trade was intra-industry,[16] and trade within the CMEA has been to a large extent complementary—manufactures for raw materials. Each of the East European economies tended to produce a broad range of products on its own. Specializa-

---

15. Organization for Economic Cooperation and Development, *Structural Adjustment and Economic Performance* (Paris, 1987), p. 273.

16. E. Plucinski, "Die Rolle des zweiggebundenen Austausches im Handel der EWG-Länder" [The role of intra-industry exchange in trade among the EEC states], *Aussenhandel*, 1987, no.4, pp. 37-41. The significantly higher percentage of intra-industry trade compared to the previous quoted study results from a higher level of aggregation in the latter study.

tion, particularly with respect to specific components of a product, was achieved only in some cases.

International trade patterns, such as intra-industry trade, can be partly explained by special features that distinguish many markets of technology products from other markets.[17] One important distinction is the existence of economies of scale, which prevail in the production process for a wide assortment of high-technology products. While most sectors of the economy that are highly dependent on raw materials, such as agriculture or mining, clearly exhibit declining or constant returns to scale, technology-intensive products such as electronics, aircraft, or telecommunications are likely to exhibit increasing returns to scale. The existence of increasing returns to scale means, loosely speaking, that production increases more than twice if all inputs are doubled. From the cost side, with an increase in production, average costs will decline. Declining unit costs are naturally linked with the production of high technology. These products require, in general, large research and development expenditures, and thus, production is governed by high fixed cost and therefore declining average costs. Because of increasing returns to scale, markets cannot be perfectly competitive. Thus, technology is often traded among a rather small number of producers in oligopolistic markets.

There were basically two reasons for the globalization and internationalization of production and trade that blossomed in the last decade. First, the expansion of markets beyond national borders was necessary to capitalize on declining average costs. Second, international cooperation and collaboration among firms allowed them to share the costs and risks of huge R&D expenditures, which are necessary to develop and commercialize new technologies.

Empirical studies have shown that the welfare gains from trade liberalization between Western industrialized countries may be two to

---

17. The so-called new trade theory tries to work out the consequences of such features as increasing returns to scale and imperfect competition for foreign trade. A survey of this work is given in E. Helpman, "Increasing Returns, Imperfect Markets, and Trade Theory," *Handbook of International Economics*, vol.1 (North Holland Amsterdam, 1984). See also Paul R. Krugman, *Rethinking International Trade* (Cambridge and London: MIT Press, 1990).

three times larger when imperfect competition is accounted for.[18] Of course, these results need not necessarily hold for liberalization of East-West trade. However, they increase the likelihood that barriers to trade, such as export controls, generated higher losses (unused profit opportunities) for Eastern Europe than was assumed in the earlier literature. To put it differently, the gains from liberalized East-West trade in the presence of imperfect competition and economies of scale can significantly exceed gains expected under perfect competition.

In the future, the East European economies will have to participate in this kind of trade. The integration into the worldwide division of labor is inevitable; this applies also to a large country like the Soviet Union. The task of integrating into the world economy has now been accepted by most East European leaders. They understand that opening the economy to the world market is a necessary (though not sufficient) condition for catching up to the level of development in the industrialized countries. The success of the newly industrialized economies in Southeast Asia strengthens the concept of an open export-oriented policy.

However, integration into the existing world market, which is dominated by the West, does have its risks and dangers. Participation in the division of labor requires that domestic industry be competitive. As explained in an earlier section, faulty policies and the inflexible organization of the economy explain to a large extent the loss of competitiveness in Eastern Europe. Assuming that the negative influence of these domestic factors will disappear or at least significantly decline with the market-oriented reforms, there still remain some particularities of the technology markets that can help or hinder the integration of Eastern Europe and the USSR into the world market.

Following are examples of how distortions, imperfections, or simply peculiarities of the market have affected the specific situation of the East European producer.

In general, Eastern Europe cannot follow the trend of sharing risks and costs of R&D by cooperating with producers in the West. As long as export controls are applied to dual-use technology, East European

---

18. See J. David Richardson, "Empirical Research on Trade Liberalization with Imperfect Competition: A Survey," *OECD Economic Studies*, 1989, no. 12, pp. 7-50.

firms are banned from joint production or cooperation with Western firms in the area of high technology. Similarly, Eastern Europe could not participate in the large Western technology programs such as the European EUREKA program. Only recently, membership of East European partners in some projects was made possible. The exchange of information about advanced technologies between East and West is also controlled. For example, export licenses are needed for US high-technology executives to talk to East bloc representatives. It was only in 1990, during Gorbachev's visit to the US, that Soviets were allowed to visit Silicon Valley for the first time.

In addition to those hindrances stemming from COCOM regulations, there are impediments that are at least as strong that have resulted from the inability of the East European side to deal with markets.

One trend that works against East European producers has been the shortening of product cycles. Traditionally, the East European economies produced at the end of the technological life cycle, when the technological product was already fairly standardized. As product cycles became shorter, prices at the end of the cycle fell more quickly. The faster new products took the place of old ones, the more expensive it became for Eastern Europe when the appropriation of product cycle technology was delayed by Western export controls. According to a study by the consulting firm McKinsey & Co., a delay of six months in marketing a product with a life cycle of five years reduces profits by one-third. Compared to this, a 50% increase in development costs decreases profits by only 3.5%.

As mentioned, integration of local markets allows firms to increase the size of their sales market and thereby take advantage of economies of scale. One attempt to do so is standardization. When there is a variety of nonstandardized, incompatible products, users and producers hesitate to buy or invest in these products. Besides reducing this hesitancy, standardization has additional advantages. The existence of a standard reduces the search cost of the user. And often, the higher the number of users of a given product, the more valuable that product becomes to its user. For example, the telephone user's benefits increase as more and more people are connected to the same telephone network. Similarly, a broad base of compatible computer systems allows users to share software and expertise. To gain the advantages of mass usage, compatibility of products is required, and this can be assured by standardization.

Often these advantages of standardization spur demand to such a level that economies of scale in production can be exploited. The de facto standard for personal computers, introduced by IBM in 1982, and the ensuing explosive development of this industry illustrate the significance of compatibility and standardization. Participation in setting standards not only gives firms a lead in information about the future of the market, but, more importantly, it allows them to take part in the actual shaping of those standards and norms. The fierce debate in the West over standards for high-definition television (HDTV) demonstrates that high profits are at stake.

Eastern Europe has hardly taken part in these international activities. To exploit the opportunities from standardization, countries must become members or participants of various international organizations in which, formally or informally, standards and norms are decided upon. In many cases, the absence from these meetings was due to restrictions on the part of the East European countries themselves.

The exchange of know-how is a vital part of the production of technology. Among firms, this exchange is by no means a simple transaction of goods for money. Frequently it is barter trade (information for information). The successful transfer of technology requires cooperation, and it is often a mutual exchange of know-how by both participants. It must be accompanied by additional advice and consultation over a period of time, and generally cannot be completed at one go. Such mutual dependence adds to the confidence of both partners that all relevant aspects of the know-how are transferred. Thus, firms look for partners with technological competence that not only take but also can give know-how.[19] In this field, East European firms can hardly compete. As suppliers of knowledge they have a bad reputation. Thus, Western firms are reluctant to provide East European firms with first-rate know-how and blueprints.

---

19. E.g., the transfer of the design of the 1-megabit chip from Toshiba to the West German firm Siemens was done in an exchange of a data bank for logic chips developed by Siemens.

Cumulative advantages are generated when the exchange of technology occurs exclusively among firms that are technological leaders. The firms initially participating in an exchange of know-how will mutually benefit from this exchange and will increase their technological superiority. On the other hand, firms that were initially excluded from this exchange cannot benefit from this division of labor and will increasingly lose contact with the technological frontier. This, however, reduces even further their chances of participating in the exchange of know-how.

Finally, one has to consider the "home market effect." Due to several rounds of liberalization and difficult negotiations among major Western producer countries, many firms of the industrialized West can operate on a large international market, which by now can be considered their home market. If scale economies prevail, as for example in the production of memory chips, the firms with a large home market have a cost advantage. The integration of Eastern Europe, i.e., the opening of the East European markets, would result in a market expansion for the Western producer who can offer lower prices due to longer production runs. In general, the East European firms cannot compete under such conditions.

These are only some features of technology trade that may be relevant to the liberalization of East-West trade. Much will depend on the behavior of policy makers in East and West. The characteristics of technology trade mentioned here obviously provide plenty of opportunities for governments to intervene, claiming, rightly or wrongly, that their actions are necessary to correct market failures. In addition, imperfect competition allows firms to earn profits in excess of their costs. At least in theory, a country can use strategic trade policy to secure a bigger share of these profits for itself. Although most economists working in the area of strategic trade warn against interventionist trade policy, the temptation to engage in such policy exists and is growing in both East and West.[20] The ongoing debate among Western

---

20. For discussions on strategic trade policy from different points of view, see Paul R. Krugman, ed., *Strategic Trade Policy and the New International Economics* (Cambridge and London: MIT Press, 1988).

countries about subsidies for aircraft production or dumping of memory chips are cases in point. When Eastern Europe integrates into the world market and the volume of East-West technology trade increases, more and more attention will be paid to the conduct of East-West trade and the role of government intervention.

In view of the strong East European tradition of interventionist policy, arguments for strategic trade are likely to be used extensively in setting up an industrial policy. One cannot deny outright the need for such a policy in Eastern Europe in the coming years; however, one has to warn of the danger that old mercantilistic thinking will reappear with a new terminology. The old bureaucracy in the East European economies could serve as a solid institutional base to support protectionistic policies. In addition, the far-reaching structural adjustments necessary for the East European economies will be used as arguments to depart from free trade.

On the other hand, if the barriers to entry into some technology markets are as high as suggested above, one must consider how they can be overcome. One proposal put forward in the recent debate suggests a stronger cooperation among the East European countries. The European Payments Union after World War II has been pointed out as a successful example of cooperation among weak economies. A kind of customs union is another concept that may be useful as a means to develop skills in a limited market and at the same time to gradually integrate into the world market. At the moment, however, the states of Eastern Europe do not seem too eager to consider such suggestions; rather, they seem to be competing against each other in a race to sign contracts with the West.

Lifting restrictions on technology flows will not suffice in helping the reforming economies. The West must, in addition, open its markets for goods from Eastern Europe, even if this competition begins to hurt. Providing access to the Western market may be the most efficient aid the West can offer. And special care should be taken to detect and dismantle any tariffs or non-tariff barriers, such as different norms, or problems in licensing technology, which could hinder the development of intra-industry trade. This kind of trade is not only a significant source of gain for all partners, but it also leads to a tight network of trade flows. Such a network will create mutual dependencies, which in turn can help to stabilize the process of integration.

Implications for Policy Makers in East and West

The question of easing curbs on East-West technology transfer is part of a more general problem: What will the reforming economies in Eastern Europe gain by participating freely in the worldwide exchange of technology?

Empirical data demonstrate clearly that, in the past, Eastern Europe hardly took part in international high-technology trade, and that it progressively lost its already weak competitiveness in this field. The institutions and incentive structures in Eastern Europe effectively hampered the development and adoption of modern technology. These system-specific limitations, which prevented Eastern Europe from utilizing the productivity-enhancing potential of technology, masked and blurred the otherwise more visible damage caused by Western export controls. The effect of these controls on the East European economies was not negligible. Deprived of high-technology imports from the West, the socialist economies had to develop their own indigenous technology, at high cost and with rather limited success. More often than not, this resulted in a costly misallocation of resources, as demonstrated by the electronics industry in the GDR.

During the transition towards a market economy, needs for technology have developed that had not existed in the old system. Financial and banking systems require rather advanced information technology, and pressing environmental problems create their own demand. At the same time, with the changing climate caused by the reforms, new private and governmental sources have become available to finance the emerging demand for technology.

In light of the damage caused by Western export controls to the old socialist system, and considering the increased role of technology in the reforming economies, it would seem logical to conclude that Eastern Europe will reap immediate and significant benefits from a liberalization of East-West technology trade. However, some caveats should be added to this proposition.

Technology is traded, to a large extent, on imperfectly competitive or monopolistic markets. The peculiarities of international technology trade may help Eastern Europe derive advantages from participation in this trade, or may hinder it.

It has now become feasible for Eastern Europe to participate in the widespread cooperation of firms at a preproduction level in sharing risks of R&D or reaching agreements on norms and standards. Taking part in these early stages of development might save Eastern Europe from remaining a victim of the ever-shortening product cycles, as it had been in the last decade. But liberalization might also provide the playing field for the West to exert its market power, to use its cost advantages from mass production that result from its large home market. The reluctance of Western companies to share know-how with technological have-nots adds to the fear that East European markets will be conquered by Western firms while domestic producers will be completely driven out of the market once governmental protection is withdrawn.

It is tempting for East and West to use interventionist policies in pursuit of the benefits of liberalized East-West trade. To prevent backsliding to mercantilistic trade policy, a joint East-West effort should be undertaken to shape the conduct of East-West trade in a way that takes into account the obvious imperfections of the technology market and the unequal market power of the players. The rules of free trade, canonized in the case of perfect competition, may not be sufficient to integrate Eastern Europe into the international division of labor.

4

Global Technology Markets and Security Issues

IGOR E. ARTEMIEV

The perception of security is changing in the post-Cold War world. Perestroika, "new thinking" in foreign policy in the USSR, and the dynamic changes in Eastern and Central Europe have created more freedom to choose political and economic systems based on expanding democratization. Success stories of privatization and deregulation in developed and developing countries, as well as the failure of the command economy model of socialism to satisfy the basic needs of the population (to say nothing of meeting the world technological challenge) have increased the appeal of the market economy model for the countries that had been preaching rigid centralization for the whole postwar period. The prevailing trend in this part of Europe at present is marketization, with varying degrees of acceptance of socialist ideas. The laws and principles of market interaction have been recognized in both international and intranational economic activity everywhere, creating a common basis for the post-Cold War world trading system.

In addition, because of the dramatic changes in Eastern Europe, the developed market economies feel that the East poses less of a threat to them. Many Western policy makers believe that these former potential adversaries will not emerge again as military threats in the foreseeable future, even in the event of a Soviet reversion to the old centralized system, however unlikely that scenario may be.

Economic issues are gaining ground in the discussions on international security. Military, political, and economic issues are increasingly interrelated, and in certain areas, like state control on exports of technology, they are interdependent. Speaking at the March 1990 CSCE Conference on Economic Cooperation in Europe, FRG chancellor Helmut Kohl called for a reappraisal of export controls:

> The European security partnership [for its part] boosts the economic partnership; it creates the political freedom necessary for the development of visible and lasting cooperation between businesses. Moreover, basic political reorientation and tangible progress towards disarmament must go hand in hand with an examination of existing export controls on both sides, for the process of increased openness and integration must not be hampered unnecessarily by trade barriers.[1]

Because of their strategic importance, markets for high-technology products and technical know-how are subject more than any other commodity market to export controls. I will begin my discussion of this issue with a short summary of world technology markets. I will then compare technology transfer practices in the command economy to the diffusion of technical know-how in the market economy. I will also analyze the various forms of compensatory relations in technology transfer used by the Council for Mutual Economic Assistance (CMEA) member countries, and how these efforts have led to the understanding that marketization will not be successful unless it is introduced into most sectors of the economy, not only in science and technology.

I will also analyze recent trends in Western export controls and much less publicized efforts by the USSR to modify its system of technology export regulations. I will conclude with a view of the state's future role, based on cooperation and mutual confidence, in technology markets in the post-Cold War world, and suggest a system of effective verification of the use of controlled items in the USSR, which

---

1. Speech by Helmut Kohl, Chancellor of the Federal Republic of Germany at the opening of the CSCE Conference on Economic Cooperation in Europe in Bonn, March 19, 1990. Official translation by the government of the Federal Republic of Germany, p. 7, Bonn, March 20, 1990.

will significantly reduce the costs of technology transactions for both sides and raise East-West confidence and cooperation.

Global Technology Markets

Technology is production-oriented, specialized knowledge necessary to turn resources into goods and services. It is *not* just scientific knowledge, nor is it just material goods in which technology is often embodied; it is *not* merely a unique service that is normally nonstorable and hard to accumulate. Technology is a product of incremental research, development, and engineering efforts, and can be mastered only if production factors are sophisticated enough to assimilate the innovation and if management and labor are motivated to overcome barriers and resistance to technological change. If an economic system resists change and lacks motivation to introduce new technology, it is doomed to technological backwardness, even if the most sophisticated means of production are available on the market.

The competitiveness of national economies is increasingly determined by technological factors, and above all by a capacity to implement innovation effectively. By virtue of its dynamism, technology is an important cause of economic instability. It has become a serious threat, particularly to the security of those countries that are not capable of adjusting quickly to changes in production. For example, at the February 1990 National Policy Review of Technological Progress in the USSR, held in Moscow, speakers emphasized that Soviet "neglect of the growing role of microelectronics, biotechnology, and resource- and energy-saving technologies had led to the omission of a whole series of breakthroughs" and had contributed to the deterioration of the level of social and economic development of the country. As a result of the neglect, the "technological gap with the industrially developed countries has been growing noticeably."[2]

---

2. Nikolai Sliunkov, quoted in A. Pokrovsky, "Chto rozhdaetsia v sporakh" [What results from debates], *Pravda*, February 20, 1990; and "Novye podkhody k nauchno-tekhnicheskoi politike" [New approaches to technological policy], *Pravda*, February 22, 1990.

Although technology is primarily an input created by economic agents for in-house use, it became a commodity as early as the 18th century. Technology markets have loomed large in recent decades, when technology started evolving more rapidly than at any other time in human history. There were many reasons for marketing technology, including the growing role in the division of labor of R&D and engineering firms, whose final product is technology; "overproduction" of technology, when a firm cannot use all inventions and innovations developed in its R&D labs and experimental plants; and commercialization of accumulated know-how, especially when a firm is planning to replace one technology with a newer, more productive one.

Technology markets, which play an important role in the creation, implementation, and diffusion of technology, are important elements of a market economy. Four kinds of international technology markets can be identified:

1. *The arm's-length transactions market* is a classic form of interaction among independent economic agents, e.g., between firms that have no equity or other ownership rights in each other's assets. Due to market imperfections and business strategy considerations, however, this market plays only a marginal role in the international transfer of technology.

2. *Internal 'markets'* were created by multinational corporations (MNCs) in order to overcome the inefficiencies of the arm's-length markets for technology. Technology transactions between parent companies and their foreign affiliates constitute a substantial part of the world technological exchange.

3. *Strategic alliances and coalitions* are quite recent phenomena, resulting from increased technological cooperation among MNCs based in different countries, and aimed at coping with rapidly evolving technologies. In the 1980s, the spread of these partnerships was spectacular, but their implications are not yet fully understood. These relationships are not intra-firm by definition, yet they lack explicit arm's-length characteristics.

4. *Government markets* play a visible role in the transfer of technology under government procurement or when the state is a party to the transaction. In this case, public good characteristics of technology are important.

Although fees, royalties, and similar payments statistics are incomplete and contradictory, these data remain the only indicators of the trends in the international technology markets. According to statistics from the Organization for Economic Cooperation and Development (OECD), the value of international payments for technology reached $13 billion by the mid-1980s, with an annual real growth rate of 2.5% (which was higher than that of trade in tangible goods or foreign direct investment).

Developed market economies are the major participants in the world technology market. Their share in technology receipts is close to 100% and in payments about 75% (see Table 1). Most Western countries are also technology importers. The US accounts for more than half of the world's registered payments for disembodied technology (its share in the global intangible technology market is twice the size of its share in world exports and much greater than its share in foreign direct investment worldwide). The US, the UK, Sweden, and Denmark (and perhaps Switzerland, although statistics are lacking) form the group of net exporters of technical know-how. All other developed market economies are net importers of technology. The South and the East are mainly importers of technology (their respective shares in world payments for technology are 20% and 5%). During the 1980s a major technology importer—Japan—has significantly diminished its deficit in technology trade. Most of Japan's receipts come from the developing countries (with Southeast Asia foremost), while other Western technology importers sell more to other countries in their own group than to other country groups.

The role of the USSR and Eastern Europe in the disembodied technology markets is negligible. The USSR's annual income from the sale of licenses to foreigners is below $300 million, and for the other East European countries it is less than $20 million. Most of the technology exported by Eastern Europe is at the "invention-conceptualization-through-pilot-plant" stage, before implementation in mass or serial production. Technologies implemented in mass or serial production predominate in the licensing purchases of the East

Table 1
Dimensions of the World Markets of Disembodied Technology (fees, royalties, and other technology receipts and payments) 1975-1985 (at 1980 prices and exchange rates)

Country/Group of countries	Receipts				Payments				Balance	
	1975 bln. dollars	%	1985 bln. dollars	%	1975 bln. dollars	%	1985 bln. dollars	%	1975 bln. dollars	1985 bln. dollars
United States	6.2	62	6.8	54	0.7	9	0.2	2	5.5	6.6
Japan	0.4	4	0.9	7	0.9	12	1.2	13	-0.5	-0.3
European Community[1]	3.1	32	4.4	35	5.4	70	7.2	75	-2.3	-2.8
Other Developed Market Economy Countries[2]	0.2	2	0.5	4	0.7	9	0.9	10	-0.5	-0.4
All Developed Market Economy Countries	9.9	100	12.6	100	7.7	100	9.5	100	2.2	3.1

[1]Including payments/receipts among EC member countries. No data for Ireland and Luxemburg.
[2]Australia, Austria, Canada, Finland, Norway, and Sweden.
Source: *OECD Science and Technology Indicators Report*, 1989, no. 3, p. 34.

European countries. Since the latter are much higher priced than the former, these countries have a deficit in technology trade, although the number of licenses sold and purchased may be almost equal (as in the case of the USSR).

All students of the subject have found that the functioning of a market based on arm's-length technology transactions is notoriously inefficient, due to imperfect competition. Some of the factors leading to market failure include the small number of participants in the market, the buyer's incomplete knowledge of the product, and the innovating firm's difficulties in appropriating the benefits of the new technology because of its public good qualities.[3] All agents in the technology market claim that they lose more than they gain in the transactions. Although some US economists feel that "technology is the most attractive bargain of the 20th century,"[4] the notion of government control or monitoring of the export of technical know-how, for the purpose of maintaining US economic competitiveness, has been actively debated in the US since the late 1970s, but the results in terms of practical policies have been quite limited.[5] Developing countries, net importers of technology, conceive of technology as a common human heritage, which should not be monopolized to extract excessive profits.[6] Nevertheless, their efforts to adopt a binding code of conduct for technology transfer, which would enhance the rights of importing countries to regulate technology trade and bring down the price, have also failed.

The USSR and the East European countries supported the idea of a

---

3. See, for example, Richard E. Caves, Harold Crookel, and J. Peter Killing, "The Imperfect Market for Technology Licenses," *Oxford Bulletin of Economics and Statistics*, August 1983, pp. 249-67.

4. Raymond Vernon, in *The Political Economy of International Technology Transfer*, ed. John R. McIntyre and Daniel S. Papp (Westport, CT: Quorum Books, 1986).

5. See Michael Boretsky, "Trends in US Technology: A Political Economist's View," *American Scientists*, January-February 1975; Jack Baranson, *Technology and the Multinationals: Corporate Strategies in a Changing World Economy* (Lexington, MA: Lexington Books, 1978); and others.

6. See, for example, D. Gezmidis, *Le transfert technologique par les firmes internationales* (Paris: OECD, 1977).

code of conduct for technology transfer, but they did not share the concept of technology as a common human heritage. This position contrasted with their approach to technology transfer within their national boundaries, which was arranged on a noncommercial basis in accordance with mandatory planning directives. Competition among enterprises did not exist, and minimization of parallel R&D was a priority.

Technology Transfer in the Command Economy

Technology diffusion in the command economies has explicit noncommercial characteristics, in contrast to the rather wide variety of commercial relationships in the market economies. For example, the R&D and engineering efforts of innovating enterprises are usually financed from the state budget and the technology they develop does not become their property. The theory of orthodox socialist political economy postulates that all benefits of an innovation are to be directly collectivized (i.e., de-privatized) like any other rent income.[7] Private property (including intellectual private property) is considered a major impediment to technology diffusion, and any commercial relationship using scientific and technological know-how is viewed as a contradiction of its public good character.[8]

Soon after the Revolution of 1917, the Decree on Workers' Control was adopted in Soviet Russia, with the aim of strengthening state authority over private enterprises. This decree abolished the notions of "commercial confidentiality" and "trade secrets," and terminated the traditional patent system that had been in practice since 1812. Later, the author's certificate became the prevailing form of legal protection of inventions. This document recognized the authorship of the inventor, but assigned all the rights to use the invention to the state. The inventor was entitled to royalties up to a certain limit whenever the invention was used by a state-owned enterprise.

---

7. See Aleksandr Anchishkin, *Nauka, tekhnika, ekonomika* [Science, technology, and economics], 2nd ed. (Moscow: Ekonomika, 1989), p. 260.

8. Ibid., p. 256.

A major reason for prohibiting commercial technology transactions in a command economy was the fear that if an innovating enterprise were allowed to appropriate the benefits of an innovation, the additional income generated would be higher than the social productivity gains (regardless of the method used to measure these gains), and would lead to excessive price increases on new technology. Mandatory planning was the only incentive imposed from above to introduce new equipment and technology, but it was mainly a negative incentive. If plans for the introduction of new technology were not fulfilled, the management could be punished or dismissed. Yet if the plan was met, the enterprise did not get sufficient compensation to recoup its investment in innovation. In addition, the innovating enterprise was obliged to spread the new technology and know-how within the industry, although it was repaid only for the overhead cost of the transfer operations (e.g., reproduction of blueprints, per diem for travel, etc.). Because technology was considered a public good and belonged not to the innovating enterprise or to an individual inventor but to the state, other enterprises had equal rights to use this know-how, and they expected that the creator would share the technology free of charge. It is evident that, under these circumstances, the innovative enterprise had no incentive to transfer technology effectively.

The noncommercial mode of internal technology transfer was used by all East European countries, although many of them preserved the traditional patent system. Nevertheless, as all (or almost all) the possibilities for applying the inventions remained within the state-owned enterprises, governments invariably played a decisive role in technology transfer, leaving a rather limited role for the market.

The noncommercial mode of technology relationship was practiced by enterprises *within* each individual economy and was also applied to technological cooperation *among* nonmarket economies in the CMEA. During the 1950s and 1960s, scientific and technological cooperation among the member countries was carried out in accordance with the principles adopted by the second session of the CMEA council in Sofia in 1949. The aim of the so-called Sofia principles was to "facilitate the flow of scientific and technological achievement among the fraternal countries in order to decrease the burden of financial obligations resulting from external technology borrowing." The recipient country had to cover nothing but the costs of reproducing and mailing the technical documentation, promising not to use it outside its national territory. The adoption of the Sofia principles was aimed at facilitating the policies of Soviet-type industrialization in the East European countries. From 1948 to 1978, the USSR reportedly transferred 35,000 sets of technical, engineering, and other documentation to the CMEA member countries and Yugoslavia, and supposedly received about 17,000 sets of similar documentation from them. Soviet experts estimated that these exchanges saved the other CMEA countries up to $20 billion, which otherwise they would supposedly have had to spend to purchase the technology on the world market.[9] (One has reason to doubt the meaningfulness of such estimates, when technology transfer is equated with mere copying and mailing of information. Technology transfer can be reduced to information dissemination only when both partners are equally developed and possess strong capabilities for the quick and effective transformation of information into productive technology. This was obviously not the case in postwar Eastern Europe.)

The concept of technology as a public good was paralleled by the CMEA countries' relationship with the developing countries. A substantial amount of technical data and know-how was transferred by the USSR and some East European countries as part of turnkey and technical assistance projects in less developed countries (LDCs). This assistance was free of charge, except for the costs of preparing and transferring the documentation. If buyers needed appropriate technology, they were usually charged only for the necessary technological modifications. Even if a significant amount of extra work for adjustment of technology was required, the recipient was likely to reimburse only the costs of additional development engineering and design, and still received the basic technology free of charge.

The only deviation from the noncommercial mode of technology diffusion in the command economies took place in their relationships with the market economies. Foreign sales and purchases of R&D or engineering products were conducted on the basis of market prices. This was the only exception to the usual noncommercial relationships

---

9. Oleg T. Bogomolov and Aleksandr N. Bykov, eds., *Strany SEV v mezhdunarodnom obmene tekhnologiei* [The CMEA countries in international technology exchange] (Moscow: Mezhdunarodnye Otnosheniia, 1986), p. 64.

in technology diffusion.[10] In order to cushion the shock of fluctuating world markets, the Soviet government introduced a system of specialized foreign trade organizations (FTOs), which functioned as intermediaries between national enterprises and foreign firms. Each of these organizations had a monopoly on exports and/or imports of a particular product, and at least one of them dealt with disembodied technology (primarily licensing and technical assistance). In East-West trade, technical know-how was treated as a commodity. Later, this came to be increasingly true for East-South and East-East trade as well.

Building a Technology Market in a Command Economy?

By the late 1960s, it became evident that exclusively noncommercial modes of technology transfer did not provide an adequate foundation for the growing number of intra-CMEA programs of cooperation in science and technology. Mutually beneficial cooperation was no longer envisaged without "strict contractual stipulation of obligations, and detailed accounting of material and intellectual endowments of the partners, in proportion to the benefits received from the cooperation."[11] In 1967, commercial forms of compensation for transactions related to scientific and technological programs were officially legalized. These compensatory relations were not market transactions, but payments to the transferors allowed them to recoup (wholly or partially) their R&D or engineering costs, and purchasers did not share the profits resulting from the use of this technology. Profits from the use of technology are the main pricing factor in the world market for arm's-length transactions. Nevertheless, even in the more recent intra-CMEA licensing agreements (the first of which was reported to be concluded in 1971), it is still unclear which criterion was more important in determining the price of technology—cost recoupment or profit participation. But if one takes into consideration

---

10. Anchishkin, *Nauka, tekhnika, ekonomika*, p. 333.

11. Bogomolov and Bykov, *Strany SEV v mezhdunarodnom obmene tekhnologiei*, p. 65.

the general nonmarket environment of CMEA economic interactions (e.g., overbureaucratization, nonconvertibility of national and collective currencies to settle claims, etc.), these licensing agreements could hardly qualify as market deals.

But even this change from noncommercial to compensatory relations in intra-CMEA technology transfer was a result of uneasy compromises between more developed members pressing for change and less developed members interested in preserving the status quo. The language of the Comprehensive Program of Socialist Economic Integration was ambiguous:

> Subject to its level and value, the output of scientific and technological research can be transferred either free of charge or with financial compensation under an agreement, provided that this transfer takes into consideration the national interests of each country, as well as the common interests of all CMEA member-countries.[12]

Although various attempts were made to expand compensatory arrangements in science and technology within CMEA, the process of internal creation and diffusion of technology remained noncommercial, and the national economic systems continued to resist technological change. By the early 1980s, the technology gap with the West had become so evident that CMEA economic security and technological vulnerability, exacerbated by economic sanctions in the aftermath of the Afghanistan invasion and martial law in Poland, loomed large at the economic summit of the CMEA member countries in 1984. In order to respond to the technological challenge of the West, the economic summit approved the outline of the Comprehensive Program of Scientific and Technological Progress (adopted in 1985), in which new directions and frameworks for cooperation and accelerated technological change were envisioned. These efforts were intended to make CMEA member countries more interdependent and to add a new dimension to economic and technological cooperation. During the

---

12. Members of the Council for Mutual Economic Assistance, *Comprehensive Program of Socialist Economic Integration of the Countries* (Moscow: CMEA Secretariat, 1971).

Table 2
Regional Composition of Technology Trade of Hungary, 1988 (percent)

	Imports to Hungary	Exports from Hungary
Developed market economies	82.4	66.5
Developing countries	—	15.6
Eastern Europe and USSR	17.6	17.9

Source: *Economic Bulletin for Europe* 41 (November 1989), p. 66.

following years, however, no major breakthroughs occurred, either in technological evolution or in the level of cooperation among the CMEA member countries. The targets of the program were far from met: cooperation encountered major difficulties, and the few actual joint R&D projects did not lead to cooperation in production, as about 75% of new technology development in these joint projects was brought to the production stage only by the country coordinating the project. Production enterprises in other member countries did not care to use the technology. At the same time, their current technology requirements were being met by imports from the West. (See Table 2, which shows that mutual transactions of CMEA members are relatively unimportant for Hungary.) The CMEA was not able to overcome its reputation as a low-technology area where countries were transferring inefficiencies rather than modern technologies and know-how.[13]

The major causes of the last comprehensive program's difficulties are well known. The program failed to create a favorable environment for the direct interaction of grassroots enterprises and research institutions involved in technology creation and diffusion, and it also failed to establish the appropriate non-mandatory government mechanisms to manage these interactions. Excessive centralization undermined the economic autonomy of the economic agents involved, and discouraged their interaction. It is clear that efforts to match elements of the new market mechanism with the old command economy's "coordination" were doomed to failure.

---

13. See Vladimir Sobell in *Technical Progress and Soviet Economic Development*, ed. Ronald Amann and Julian Cooper (Oxford: Basil Blackwell, 1986).

Within a command economy, there are many difficulties in creating international cooperation schemes that embrace all stages of the innovation cycle (i.e., research-development-manufacturing-marketing). State regulatory measures are limited to assigning each supplier directly to a consumer, thus ensuring that a manufactured product can be assembled using the parts and inputs produced by other plants and industries. In such an environment, there is no room for joint projects or joint ventures initiated from below. This is all the more true if the enterprises are supervised by different industrial ministries.

The task of maintaining joint technological cooperation is even more difficult if the partners belong to different national mandatory planning systems. As the CMEA had neither a supranational mandatory planning center nor a common allocations fund, cooperation could be organized only on the basis of a very complicated system of state-coordinated activities at all levels of economic management in order to push all the participants into operational projects. But enterprises and research units deprived of economic autonomy and functioning in a noncompetitive environment lack any incentive to cooperate. Thus, efforts to develop technological cooperation by relying on activities coordinated by state agencies proved to be totally inconsistent. They led only to the intensification of bureaucratic procedures and to the division of various stages of cooperation. Many programs adopted at the senior level remained only declarations of intent, unequipped with the proper means for their fulfillment.

Nurturing productive technological cooperation requires that enterprises and research units have full-fledged economic autonomy, and systemic changes are necessary to motivate enterprises to implement new technology instead of resisting it. It is not enough to proclaim economic autonomy by law, nor is it sufficient to adopt another program of cooperation (however "comprehensive" it might be). A wholly new model of interaction is required—one that would give economic incentives to the enterprises and research institutes to establish direct ties with each other. This new model of cooperation should be based on a more efficient division of labor among the basic units of the East European countries. If this principle were followed, trade between industries would, over time, give way to trade *within* industries, and this would promote a more rational exchange of components, supplies, and inputs. In this environment, technology transfer on the basis of licenses and contracts would increase.

New Soviet Policies for Technology Transfer

The late 1980s witnessed the further opening of the Soviet economy, spurred by the decentralization of foreign trade and the introduction of flexible and transparent government regulations in external economic relations, including technology transfer. Since April 1, 1989, the international business environment in the country has changed dramatically: all state-owned enterprises and privately owned production cooperatives now have the right to trade directly with foreign partners.[14] Their access to external markets is now based on the principles of self-financing and currency self-reliance, i.e., neither the state nor any other Soviet entity can be held responsible for the enterprises' liabilities.

Since August 1986, the government has been gradually and consistently reforming the foreign trade sector, formerly based on the state monopoly of foreign trade (which had been interpreted for many decades as the monopoly of a particular government agency—the Ministry of Foreign Trade). First, the right to export and import was extended to a handful of branch ministries and state-owned enterprises. Later this right was granted to all of them, as well as to cooperatives involved in production activities. The growing role of republics and localities in Soviet foreign economic activity has been another important development of the past two years. Local authorities are challenging the exclusive right of the central government to control foreign trade. The resulting confrontation is to be resolved by a new union agreement to be worked out in the near future.

This liberalization embraced all forms of export-import operations, including technology transfer contracts. From 1962 to 1986, the FTO Licensintorg was the sole channel for Soviet disembodied technology exports, but now Soviet enterprises and cooperatives can either sell their technology directly or choose Licensintorg or any other FTO

---

14. Government Directive No. 1405, *Ekonomicheskaia Gazeta*, December 2, 1988.

to represent them abroad. Technology purchases have also been decentralized. Enterprises can use export proceeds to buy capital goods and know-how directly from foreign firms, or they can acquire tangible or intangible technology through specialized FTOs or firms servicing various industrial branches.

The new law on inventions, which was adopted in May 1991 by the USSR Supreme Soviet, contains substantial changes in the Soviet system of intellectual property protection.[15] The author's certificate, the prevailing form of legal protection of inventions in the USSR, was replaced by the patent, which assigns the exclusive right to use an invention to a definite patent holder, who could be the inventor, the enterprise, or the cooperative where the inventor is employed or to which he or she has assigned his or her invention. An invention can also be transferred to the USSR State Patent Fund, which then becomes the legal person responsible for state-owned inventions. The new law will recognize that a patent and all its proprietary rights are commodities, and it will aim at creating an internal market for licenses. The proliferation of licensing and wholesale trade in capital goods on the internal market will, it is hoped, motivate Soviet innovators to transfer disembodied and embodied technology on commercial terms to other enterprises and foreign firms.

Such a market was not needed under the command economy, when transfers of technology were centrally planned and took place basically free of charge for the transferee. Traditionally, the government stimulated overseas licensing through fiscal instruments. Since the early 1970s, Soviet enterprises have been entitled to keep up to 80% of the licensing receipts on their foreign currency retention accounts, but could keep only up to 50% of their proceeds from exports of tangible goods.

One of the basic principles of Soviet technology policy is the enhancement of the nation's technological capabilities. The USSR has the highest ratio of R&D expenditures to GNP in the world. Nevertheless, the output indicators for Soviet science and technology are not commensurate with the inputs. Soviet exporters play a negligible role in the world's high-technology markets. Soviet exports are dominated by raw materials and low-technology manufactures.

---

15. The text of the law was published in *Izvestiia*, June 14, 1991.

Since the 1960s, the USSR has pursued a policy of complementing the development of national technological capabilities with the importation of scientific and technical know-how, machinery, and equipment. Soviet FTOs have purchased licenses and turnkey plants. Also, the current tariff law (as well as the new version thereof) is aimed at the stimulation of technology imports. Machinery, equipment, and in particular high-technology capital goods are subjected to lower tariffs than are other manufactured goods.

In January 1987, another form of technology transfer to the USSR was added when a decree was issued allowing the establishment of joint ventures with foreign firms on Soviet territory. One of the aims of the decree was the enhancement of technology transfer.[16] Since its enactment, the decree on joint ventures has been amended: the 49% ceiling on the foreign partner's equity share has been removed. The foreign partner may now take on a significant role in the management of the joint venture. All key decisions are made on the basis of a consensus of the firm's board of directors. Joint venture income is subject to a 30% tax after a two-year tax holiday starting from the date of the first declared profit. No provision was made for taxing the fees and royalty payments made by a joint venture to its foreign parent company.

In many cases, government policies to acquire technology have proven to be effective only when used selectively in specific branches and with respect to specific channels of transfer. Across-the-board measures have proven to be insufficient or counterproductive.[17] It can be expected that Soviet government policies for technology transfer will aim at more expeditious assistance to enterprises and cooperatives, and at better assimilation of imported technologies through various selective measures.

Even with such measures, however, it has become clear that it is impossible to bridge the technology gap with Western countries and increase the competitiveness of Soviet exports without a fundamental

---

16. The USSR Council of Ministers Decree on the establishment and operation of joint ventures on Soviet territory with the participation of Soviet organizations and firms from capitalist and developing countries. January 13, 1987.

17. See, for example, Farok Contractor, *Licensing in International Strategy: A Guide for Planning and Negotiations* (Westport, CT: Quorum Books, 1985).

reform of the Soviet economy and the creation of product, technology, capital, and labor markets. In 1988, the legal basis for a technology market was established, and economic autonomy (self-accounting) for research institutes was introduced. To create a new economic system responsive to technological change, a new conceptual model of technological development is being discussed at forums such as the abovementioned National Policy Review of Technological Progress.[18]

Today the market for products of intellectual efforts (software, management services, business education, engineering, etc.) is probably the only growing macroeconomic market in the USSR. This is so because the consumer product market has collapsed, the building of a wholesale market for producer goods has been delayed, and legal capital and labor markets are lacking entirely.

Recent Trends in Western and East European Controls of Technology Exports

A country's export control policy is normally influenced by considerations of national security, foreign policy, and the need to alleviate shortages. The first two considerations have played by far the greater role since the late 1940s, when the developed market economies established COCOM, and trade controls (in particular in high-technology goods and know-how) became an important instrument of their national security and foreign policies.[19] During the last decade, the COCOM reach expanded geographically to include new members (Spain and Australia) and to impose its requirements on several neutral and some developing countries (Switzerland, Sweden, Austria, Finland, South Korea, and Singapore).

The conceptual change in the US, manifested in the attempt to

---

18. See the text of the new concept of management of technological progress in the USSR in "NTP i perestroika" [Technological progress and perestroika], *Ekonomika i Zhizn'*, no. 5 (January 1990), pp. 10-11.

19. See, for example, Gary K. Bertsch, ed., *Controlling East-West Trade and Technology Transfer: Power, Politics and Policies* (Durham and London: Duke University Press, 1988).

integrate the Defense Department's military-critical technologies list into the official Commodity Control List, has led to the expansion of export controls. Further, on a micro-level, private corporations have been encouraged to create a system of self-control (in addition to the national systems and COCOM) that is aimed at both exports and sales to internal markets and is extended to foreign affiliates as well. In addition, US export control regulations are valid for foreign affiliates of US multinational corporations. This extraterritorial application of US laws and regulations is a controversial, widely debated issue among Western countries.[20] In addition, during the 1980s, the US government expanded controls to include scientific exchanges. Special customs control programs prosecuted violators under criminal law and kept "gray lists" of firms with "suspicious activities."

Many specialists, however, have challenged US government estimates of the effectiveness of export controls.[21] Leaving aside the issue of cost, an East European observer is struck by the detrimental effects of Western export controls on East-West trade and technological cooperation in areas not necessarily linked to high technology. Many Western—especially US—firms have concluded that it would be too burdensome to deal with the USSR and East European countries because of the omnipresence and influence of export control authorities. The concept of "non-strategic trade," introduced by the Reagan administration to offset criticism of export controls, did little to expand East-West economic cooperation.

It is difficult to judge whether the USSR had also tried to pursue similar export control policies. But in the command economy there were no technical obstacles to pursuing national security and foreign policy objectives or to seeking to overcome short-supply bottlenecks. Because all foreign trade transactions were approved by a government

---

20. William Finan, *Estimates of Direct Economic Costs Associated with US National Security Controls* (Washington, DC: Quick, Finan and Associates, 1986), p. 12.

21. Committee on Science, Engineering, and Public Policy of the National Academy of Sciences, the National Academy of Engineering, and the Institute of Medicine, *Balancing the National Interest: US National Security Export Controls and Global Economic Competition* (Washington, DC: National Academy Press, 1987); *Assessing the Effects of Technology Transfer on US/Western Security--A Defense Perspective* (Washington, DC: Department of Defense, 1985).

body (the Ministry of Foreign Trade), all exports are assumed to have been controlled according to the criteria of national security, foreign policy, short-supply, or other considerations.

With the decentralization of external economic activity in 1989, however, thousands of grassroots enterprises and private cooperatives started to trade directly with their foreign partners. The state then turned to a licensing system, created by the Foreign Trade Decentralization Order No. 203 of March 7, 1989, which established a list of 27 product categories subject to licensing. In late 1989, by special decree of the Supreme Soviet, the list was expanded to include "all consumer goods and main inputs for production thereof." By mid-1990, the list included many major export items of the USSR: fuels, raw materials, basic chemicals, food, iron, steel, non-ferrous metals, and consumer goods. Machinery and equipment, as well as most manufactured products, were not included. The only high-technology items added to the list were "Soviet inventions and other results of scientific and technical activities."

The inclusion of disembodied technology in the list of regulated export items is puzzling if one considers the nationwide effort to expand exports and upgrade technological quality.[22] And because the supply of disembodied technology is inexhaustible, one must conclude that the inclusion of inventions and other scientific and technical achievements in the list are due to economic reasons other than short supply, or to foreign policy or national security reasons.

Selling disembodied technology abroad may incur substantial opportunity costs on the part of the licensor. For example, license fees and royalties may be significantly smaller than aggregate profits resulting from the licensor's independent use of innovation in the licensee's country (either through export of the final product of the improved technology from the innovator's home country or sale from a foreign production facility). But major requisites for both options are the implementation of the technology in the innovator's own production and the subsequent overseas marketing of the final product

---

22. The Soviet press, however, is full of critical reports about a new breed of "egoistic businessman" ready to sell anything at hand in a rush for dollars that is damaging to the internal supply situation. See, for example, Gennadii Iastrebtsov, "Rasprodazha" [Sale], *Pravda*, February 14, 1990.

manufactured on the basis of technological improvements. It is doubtful that the regulation of disembodied technology exports administered by the State Committee for Science and Technology (GKNT) is capable of motivating industry to speed up the implementation of inventions by Soviet production enterprises or to facilitate foreign marketing of resultant products. This task is complicated by the fact that the inventor and the potential industrial user are often under the supervision of different ministries or public bodies.

Motivation to innovate and export among Soviet enterprises in the manufacturing sector varies widely. In the present Soviet economic system, additional regulation of disembodied technology exports, enacted to decrease the opportunity cost of licensing exports, would probably only discourage the sale of Soviet inventions and know-how, thus depriving Soviet research institutes and pilot plants of their only source of foreign currency earnings. It would make more sense for export administrators to facilitate the implementation and marketing of products resulting from Soviet research, if economic reform increases the industry's responsiveness to technological change.

The modest amount of Soviet technology exports precludes the opportunity to use them as a tool of foreign policy. Nevertheless, it might be reasonable to create some kind of export control system in the USSR for the sake of putting something on the negotiating table when (and if) Western export control restrictions are discussed within the context of dialogue and negotiations between the USSR and the US or the USSR and other Western countries. The legalization and transparency of an export control system in the countries that have not yet entered into negotiations with COCOM can give them some leverage. This issue was addressed in Chancellor Kohl's proposal, quoted earlier, to examine existing export controls on *both* sides.

Conventional wisdom suggests that a country should establish export controls only in areas where it possesses technological superiority. There are several areas, closely related to the military-industrial complex in the USSR, in which state regulation of technology exports might be justified on these grounds. But a system of export controls and secrecy could also be maintained to hide the magnitude of technological lags. Precise knowledge of technological leads has strategic importance for the West and goes far beyond mere curiosity, especially if Eastern Europe continues to be viewed as a "potential adversary." Thus, the objectives of technology export regulations in Eastern Europe might be different from those of the West.

The Future Role of States in Global Technology Markets

The main perceptions of the East, West, and South concerning each other's technology trade regulations may be summarized as follows:

East to West: Perceptions have changed, from outright rejection to realistic acceptance of Western export controls. A new challenge for the USSR is that some East European countries may come under the jurisdiction of COCOM.

West to East: The West is not worried about East European technology controls (where they exist) and wants to know the magnitude of its own technology leads. It is becoming easier and easier for the West to gain access to East European scientific brainpower, and the West does not expect to get ready-to-use technology from Eastern Europe. The West emphasizes the protection of intellectual property rights.

East to South: Eastern Europe is concerned by the growing role of COCOM among newly industrialized countries (NICs).

West to South: Of benefit to the industrialized West is the fact that many LDCs have turned from strict regulations on technology imports (e.g., the unbundling of technology packages) to the creation of incentives for foreign capital and technology transfer. Here again, the West is concerned with the protection of intellectual property rights.

South to East: The South wants concessions in technology assistance and aid.

South to West: The developing countries' demands for appropriate technology were abandoned, along with the concept of the new international economic order. They want technology at lower costs, technical assistance and aid, and lower tariffs and non-tariff barriers for their products.

Experts in all countries increasingly realize the significant costs of export control regulations. These costs may be economic, expressed in millions of national currency units needed for maintaining a bureaucracy to enforce the regulations, in numbers of commercial contracts for which licenses were not issued or were revoked, or in estimates of the trade and technology cooperation projects abandoned due to the existence of insurmountable barriers. But the political costs are even greater, since the use of technology export controls creates an atmosphere of mistrust and conflict, as argued by British analyst Stuart Macdonald:

> In its basic form the export control issue is so esoteric and remote from real life that all of its inherent costs are probably low. But extend export controls to dual-use technology, and include information as the most valuable element of technology, and export controls become a political opportunity to take an interest in all manners of diverse activities. Wide influence and the means to act in the name of national security are powerful political weapons whose use in the hands of the obsessively dedicated threatens the health of a democratic society.[23]

Are technology controls necessary in the new post-Cold War international environment? Will the rationale for restricting high-technology sales to foreign destinations in the 1990s remain the same as in the previous decades? The international security experts see the issue from the point of view of the transferee's military strength. An economist may see the same issue from the point of view of the transferee's economic weakness. There were perhaps strong reasons for the West to control technology exports when the traditional

---

23. Stuart Macdonald, "Benefits Beyond Belief—Assessing the Cost of United States Export Controls" (Paper prepared for the Center for Economic Policy Research Workshop on Economic Aspects of International Security, London, 1988), p. 30.

command economy was in place in Eastern Europe. That system was capable of mobilizing all resources required by the military-industrial complex, without serious regard to the costs of these efforts, and without being subjected to public scrutiny as to the effectiveness of the military expenditures. There might also be strong reasons for the West to strictly control technology exports to the USSR in the future, when and if a competitive market economy is created there. But during this period of transition, the system is "neither plan nor market," in the words of Hungarian economist Tamas Bauer.[24] When the negative aspects of both prevail, it is naive to view as a major threat an economy (including the military-industrial complex thereof) that is neither motivated to introduce and quickly assimilate sophisticated dual-use technologies nor capable of doing so. But the Western perception of the military-political threat has not yet evaporated from the minds of decision makers. Western governments and decision makers are not yet convinced that new political and economic realities prevail or that COCOM and export controls are irrelevant or at least obsolete.

In the foreseeable future, or as long as there is East-West military-political competition, it appears that global technology markets will be condemned to function under export control regulations. Nevertheless, within the context of traditional export controls, a new framework may be created to achieve a standstill and subsequent rollback of the restrictive measures. This framework could be created by a series of confidence-building measures (CBMs).

First, there is a clear need for East-West dialogue and cooperation on the issue of export controls. The Conference on Economic Cooperation in Europe, held in Bonn in March 1990, started this dialogue. Now it should be maintained by the Conference on Security and Cooperation in Europe (CSCE). The Economic Commission for Europe, which is already involved in analyzing issues of conversion from military to civilian production in the member countries, can also make valuable contributions to the dialogue on technology export controls.

Second, there is a need for a substantial reduction of the COCOM and national export control lists, and this has, in fact, taken place. In

---

24. Tamas Bauer, "Privatizing and Marketizing Socialism," *Annals of the American Academy of Political and Social Science* 507 (January 1990), p. 111.

early May 1990, President Bush announced a complete overhaul of the current export control lists and recommended that COCOM develop "a new core list of goods and technologies that is shorter and less restrictive than the present list" by the end of 1990. The June 1990 session of COCOM approved this proposal to substantially reduce the list of controlled goods, particularly in the area of computers, telecommunications equipment, and machine tools.[25] Nevertheless, more favorable treatment was accorded to certain East European countries ready to adopt COCOM-approved safeguards against diversion of controlled goods and technologies to proscribed destinations (the USSR in particular) and unauthorized end-users (i.e., military-producing enterprises).

Third, more transparency of unilateral and multilateral regulations of technology exports will establish mutual trust. Nobody is likely to question the validity of limited control on technologies closely related to weapons production. At the same time, East-West economic cooperation and technology transfer can expand only under predictable, transparent, and reliable state regulations.

Fourth, COCOM verification regimes, which provide for verification procedures in the importing countries to check whether the equipment and technology are in place and used as stipulated in the export license, might be extended to or created for countries, including so-called potential adversaries, that have not yet participated in the procedures. At the end of 1989, Soviet Foreign Minister Eduard Shevardnadze offered to accept on-site inspections on Soviet territory of equipment and technology imported from COCOM member countries in order to verify that the technology was not being used for military purposes. Soviet readiness to receive the COCOM inspectors was repeated in February 1990.[26]

Fifth, ad hoc group of experts should be formed to study restrictions on exports to the countries of Eastern Europe. This has been proposed by the USSR.

---

25. A subsequent meeting of COCOM, scheduled for February 1991, has been postponed.

26. "My gotovy priniat' inspektorov KOKOM" [We are ready to accept COCOM inspectors], *Izvestiia*, February 15, 1990.

Verification Regimes in Export Controls

Verification is not a novelty in the export control practices of the Western countries. The US government has made extensive use of "import certification and delivery verification" (IC/DV) procedures in order to prevent re-export of controlled items from the countries of destination to "potential adversaries." These procedures consist of notification to US authorities by the importer of the receipt of the commodity exported from the United States under the individual license, and subsequent periodic on-site inspections by designated US agents, who are normally assigned to the diplomatic corps in the importing country, in order to verify that the item (most likely a piece of equipment) is at the place stipulated in the license. Because the recipients have been allies or neutral countries, the emphasis of verification was more on assuring that the item was not removed from the permitted spot, rather than verifying its usage.

The involvement of "potential adversaries" in export control procedures will not dramatically modify the objectives and functions of the verification activities. The governments of Hungary and Poland have negotiated with COCOM guarantees similar to those granted by non-COCOM countries brought into Western export control practices in the 1980s. But verification's objectives will be the same: to prevent re-export of the controlled item to the ultimate "potential adversary," i.e., the Soviet Union. The assumption of the current export control scheme is that, once the technology falls into the hands of the Soviets, it is likely to be used for the enhancement of Soviet military capabilities. Such an assumption turns any transfer of controlled items into a threat to the national security of the Western countries. Hence the eastbound flow of technology will continue to be strictly controlled, and many high-technology shipments will be prohibited if there is a chance that they may be appropriated by Soviet agents.

The extent of the export controls may be modified if the ultimate "potential adversary" is ready to cooperate in the export procedures. The current assumption that the transfer of any controlled item to the Soviets constitutes a security threat to the Western alliance may be reduced to the assumption that the transfer of any controlled item to the Soviet *military-industrial complex* constitutes a security threat to the Western alliance. With the latter assumption, the focus of the export controls will shift from the issue of *prevention of export or re-exports to the USSR* to the issue of *exclusively civilian use of the controlled technology within the USSR*. If there is no substantial spillover of the imported technology from the civilian sector to the military sector of the USSR, there is no reason for the exporting countries to maintain security controls. But the active participation of the Soviet authorities is essential to limit the application of a sensitive item only to the civilian sector. And the existing IC/DV measures will hardly be sufficient to guarantee nonmilitary use of the imported technology, because they are focused more on transfer (and retransfer) mechanisms than on the application and diffusion of high-technology items.

It must be emphasized that controlling the application of machine tools or equipment—and even more of intangible (disembodied) technology—is an extremely difficult task. It is complicated by the fact that these controls would be applied by a foreign government in an alien political and economic environment that historically has been unfriendly. New concepts of controls are needed. These controls may be conceived in an atmosphere of trust and cooperation, but they must be suitable for "all-weather" conditions, including changes in overall relations due to inevitable fluctuations in East-West relations. Export control cooperation, if designed only for "fair weather" in East-West relations, will hardly last long.

Any verification regime in export controls must be based on the receiving side's guarantee of strictly civilian application of the imported technology. A precedent for guaranteeing nonmilitary use of transferred technology may be found in the Charter of the International Atomic Energy Agency adopted in the 1950s.

In designing an effective model of East-West cooperation in export controls, it is necessary to analyze the treatment of verification in US-Soviet military agreements, particularly in the negotiations on intermediate-range nuclear forces (INF). In the INF Treaty a major breakthrough took place: for the first time a comprehensive verification system was created, comprised of detailed exchanges of information, routine inspections and intrusive verification methods, e.g., short-notice (36 hours) inspections, and permanent or temporary observation and monitoring of production plants.

Such techniques of verification are very effective[27] and can be used in East-West export control cooperation. Their implementation will permit the fulfillment of the following objectives: 1) the detection of violations of the licensed use of technology; 2) the deterrence of violations, since verification increases the risk of detection; 3) the building of confidence by using a more stable and predictable framework for East-West technological cooperation; and 4) the decrease in cost of East-West cooperation in technology due to a reduction in overhead for export control on both sides. Verification in the USSR will give the Western countries an opportunity to obtain firsthand data on technology in the Soviet Union, thus cutting extensive costs on foreign availability research and intelligence operations, now performed with Western technical and human resources. The Soviet enterprises will gain an opportunity to purchase required technology at the market price without the extra costs formerly associated with export control mechanisms.

There are two major differences between future export control verification regimes and their prototypes in the military field: 1) the problem of reciprocity, and 2) uncontrollable diffusion of disembodied technologies.

Reciprocity is an important factor in the verification regimes of recent East-West military treaties. Because of the structure of its military forces, the Soviet Union has made two to three times less frequent use of verification measures (e.g., on-site inspections) than the NATO countries, and the notion of "asymmetrical verification" has been recognized as a modus vivendi in military treaties. However, these principles could hardly be applied in the verification regime of export controls. The number of Soviet ready-to-implement dual-use technologies of interest to the West is substantially smaller than the number of Western dual-use technologies of interest to the USSR. This magnitude of asymmetry leaves practically no room for reciprocity in verification regimes. All verification measures, therefore, are likely to be implemented on Soviet territory by COCOM member countries,

---

27. See a recent analysis by Volker Kisnzendorf, *Verification in Conventional Arms Control*, International Institute for Strategic Studies Adelphi Papers, no. 245. (London, 1989).

who will be required to guarantee confidentiality in matters of commercial or proprietary information acquired by their inspectors.

The second difference involves control of the diffusion of technological knowledge. If tangible items can be relatively easily controlled by modern technical devices (e.g., signal-emitting tags, monitoring instruments, or short-notice inspections), any effort to control diffusion of knowledge is doomed to failure. This problem is another disadvantage for the Soviet Union, since it maintains substantial leads in the pre-production stages of innovation cycles. Diffusion of this kind of knowledge is quite difficult to control, in comparison with more tangible embodied technology, in which the West is the leader. This difference undermines the notion of reciprocity in export control verification.

Making the decontrol of the COCOM industrial list dependent on the degree of reallocation of resources from the military to civilian sector is theoretically strong, but practically hard to verify. It is enough to recall the discussion among Soviet economists in April 1990 on the magnitude of military expenditures in the USSR to understand the difficulty involved in evaluating and reallocating military production resources to civilian needs. The abolition of items on the COCOM lists, proposed by Shevardnadze as a reciprocal concession for the Soviet acceptance of export control verification inspections, is hardly feasible, due to the logical contradiction of this proposal: inspections will have nothing to verify if the lists separating controlled items from easily accessible goods are abolished. The idea of more transparency and openness in export technology regulations and administration is therefore very timely, and practical measures aimed at more exchanges of information between the two sides are certainly an important requisite for dialogue and cooperation in export controls.

The basic criterion for the effectiveness of export controls of dual-use technology is the verifiability of its nonmilitary use. This will lead to liberalization of export control regulations on items that can be verified for exclusive civilian use in the USSR, while the old system of restrictions is likely to be preserved on items that cannot be effectively verified. Both policies start from the premise that there is a list (or lists) of commodities to be controlled for security reasons. A warning must be made against overreliance on the old system of export restrictions and underutilization of the new approach, i.e., verifying nonmilitary use.

In recent East-West military treaties, the establishment of "parameters of acceptable uncertainty" was found to be indispensable, because even with the most intrusive system, perfect, 100% verification cannot be achieved. This is even more true in verifying whether a production line is manufacturing products for military or civilian use. It seems impossible to determine a priori at the production stage the final use of the many products of a general-purpose nature (e.g., food, consumer products, transportation vehicles, air carriers). Parameters of acceptable uncertainty must be established in the verification of nonmilitary use, and the system of tight controls must be restricted to only a narrow group of items directly related to the production of offensive weapons. These items will be excluded from export control verification, due to real security concerns on both sides: from the exporting side, these concerns are based on efforts to maintain technology leads in modern weaponry; from the importing side, secrecy is needed to hide the magnitude of technology (and other) lags, which would have become evident as a result of the intrusive verification methods. But areas directly related to offensive weapon production seem to be rather limited, and there is substantial room for Soviet-Western cooperation in export control verification on Soviet territory.

Conclusion

The 1990s will be an era of dynamic change in the geopolitical and world economic situation. The readiness of national decision makers to cope with the rapid pace of change is indispensable for adequately adjusting to it.

Global technology markets form an important part of the world market economy. The Soviet Union and Eastern Europe increasingly realize that the command economy cannot meet the modern technological challenges, and that full-fledged internal markets for technology must be created. Measures adopted by the Soviet Union and Eastern Europe on technology transfer and on marketizing their economies vary by country and by industry. At the moment, these measures tend to be insufficient and inconsistent. For example, in the draft of the new conceptual model of technological development in the USSR, the role of the market mechanism is visibly underestimated.[28] As long as there is no internal market for technology, the East European countries will not be able to catch up technologically; they will fall behind in military-related technologies and continue to do so in the perestroika period, especially given that military expenditures are likely to be curtailed and certain civilian programs will receive higher national priority.

Nevertheless, many Western policy makers approach export control issues in the old tradition of the Cold War military rivalry. The COCOM system, even in its "shortened-list" version, damages East-West economic and technological cooperation and causes mistrust, due to its unpredictable and reversible decisions. The East European countries are opening up in order to become part of the world economy in which all economic agents may operate consistently, but multinational corporations can function most efficiently in these countries if all their overseas affiliates can use compatible technology. Creation of technology "appropriate" for export control requirements is costly, and such regulations will deter MNC investment in, and commercial relations with, the East European countries.

There is a need for more confidence-building measures in export regulations, and for an export administration based on cooperation. The Iron Curtain is being lifted: the Berlin wall—the most obvious example thereof—has been dismantled. Among the walls still in place, the export control "wall" dominates the landscape of East-West economic and technological relations. It is time to dismantle another Cold War wall of mistrust and isolation. An effective verification regime may be the best tool to use in starting that process.

---

28. "NTP i perestroika."

5

US Export Controls in Transition: Implications of the New Security Environment

GARY K. BERTSCH
STEVEN ELLIOTT-GOWER

Introduction

It is difficult to begin a discussion on any aspect of East-West relations these days without sounding trite. We all know that the 1990s will be very different from preceding decades. The fundamental security environment has changed so as to undermine, if not invalidate, many of the premises of US and Western policies towards the communist and "post-communist" East. Among those policies is the US export control policy. The basic question we address here is, what are the implications of the new security environment for US export control policy?

In this chapter we outline the existing system of US and multilateral export controls, describe what we take to be the most relevant fundamentals of the new security environment, and then consider their implications for US export control policy. We argue that, because the new security environment is so very different from that which

---

This chapter reflects preliminary thinking related to a new, four-year research project of the Center for East-West Trade Policy entitled "Making the Transition: Export Controls in the 1990s and Beyond." The project is partly funded by The Pew Charitable Trusts.

preceded it, the implications are profound. Indeed, because the very *concept* of security has evolved, lingering Cold War notions of strategic export controls are clearly outdated and largely redundant. We conclude with some suggestions about what a future export control or "technology management" system might look like, and raise some questions that warrant further investigation. Our aim is certainly not to develop a blueprint for US export control policy, but rather to raise some of the fundamental issues that policy makers will face in the 1990s.

The Cold War System of Export Controls

Since the end of World War II, the West has maintained strict controls on "strategic" exports to communist and, now, "post-communist" destinations, including the Soviet Union, Central and Eastern Europe, and the People's Republic of China. These controls have been coordinated since 1950 through the 17-member Coordinating Committee for Multilateral Export Controls (COCOM), an informal, low-profile, non-treaty organization based in an annex of the US Embassy in Paris.[1] COCOM maintains three control lists: one for munitions, one for nuclear-related materials and technologies, and, most controversially, one for high-technology industrial items thought to have significant military utility. Applications to export items above a certain technological level on the lists must be submitted to COCOM for unanimous approval. Items below this "Administrative Exception Notes" (AEN) level may be exported at the "national discretion" of the individual governments.

COCOM controls are given force of law and implemented through domestic legislation in each member country. In the United States, the controls have been enacted through the Arms Export Control Act (1976), the Atomic Energy Act (1954),[2] and the much-debated Export Administration Act (1979) as amended. Controls on dual-use

---

1. For more on COCOM, see Michael Mastanduno, "What Is COCOM And How Does It Work," in *The Post-Containment Handbook: Key Issues in US-Soviet Economic Relations*, ed. Robert Cullen (Boulder, CO: Westview, 1990), pp. 75-105.

2. As amended by the Nuclear Non-Proliferation Act (1978).

industrial items are implemented through the labyrinthine Export Administration Regulations (EAR), which are administered by the Department of Commerce, with input from the Departments of State and Defense, and occasionally from the Department of Energy, the Central Intelligence Agency, and the Nuclear Regulatory Commission.

In addition to national security controls directed at communist and post-communist countries, the United States imposes a host of controls directed at non-communist (as well as communist) countries for "foreign policy" reasons.[3] These relate to the rather broadly defined issues of human rights, anti-terrorism, regional stability, nuclear nonproliferation, missile technology, and chemical and biological warfare.[4] Such foreign policy controls have been targeted at countries such as Libya, Cambodia, Cuba, Iran, Iraq, North Korea, South Africa, the Soviet Union, Syria, Vietnam, and PDR Yemen. Most US foreign policy controls are unilateral, although there are multilateral controls on missile technology,[5] chemical[6] and biological agents, and, of course, nuclear weapons materials and technologies.

---

3. The United States also controls certain raw materials, notably petroleum, for reasons of domestic "short supply."

4. US Department of Commerce, Bureau of Export Administration, *Export Administration Annual Report FY 1989* (Washington, DC: US Government Printing Office, 1990), pp. 35-49.

5. The Missile Technology Control Regime (MTCR) was established by seven countries in July 1987. The original members were the United States, Canada, the United Kingdom, France, the Federal Republic of Germany, Italy, and Japan. As of summer 1990, Australia, Belgium, Luxembourg, the Netherlands, and Spain had joined or were in the process of joining the MTCR. For more information, see US General Accounting Office, *Arms Control: US Efforts to Control the Transfer of Nuclear-Capable Missile Technology*, GAO/NSIAD-90-176 (Washington, DC, 1990); and Frederick J. Hollinger, "The Missile Technology Control Regime: A Major New Arms Control Achievement," in *World Military Expenditures and Arms Transfers: 1987*, ed. US Arms Control and Disarmament Agency (Washington, DC, 1988), p. 25.

6. The "Australia Group," established in 1984, consists of the United States, the EC countries, the European neutrals, Australia, New Zealand, Japan, and Canada. It has an informal "warning list" of over 50 chemical warfare precursors and a "core list" of about nine. The group targets Iran, Iraq, Libya, and Syria. "US Efforts Against the Spread of Chemical Weapons," Statement by Reginald Bartholomew, Under Secretary for Security Assistance, Science, and Technology, before the Subcommittee on International Finance and Monetary Policy of the Senate Committee on Banking, Housing and Urban Affairs, June 22, 1989, *US Department of State: Current Policy*, no. 1188.

The raison d'être of COCOM and related US national security controls was to contain Soviet expansionism. The underlying premise was that the quantitative superiority of the Warsaw Treaty Organization (WTO) in conventional forces could be offset by protecting NATO's qualitative (i.e., technological) superiority. This made sense in a world in which the WTO *existed*, had military designs on Western Europe, and possessed a quantitative advantage; and in which the West (and especially the United States) had a near-monopoly on high technology, and could therefore more easily control its supply. Forty years later, the world—the security environment—is very different.

Although highly controversial, economically costly, and sometimes politically expedient, the raison d'être and premises of certain foreign policy controls remain valid. Arguably, what may have changed is the nomenclature: yesterday's foreign policy controls may be tomorrow's national security controls.[7]

Fundamentals of the New Security Environment

To talk about the fundamentals—the essential components—of the new security environment perhaps involves a certain intellectual arrogance or foolishness, after the swift and largely unpredicted upheavals in the communist world in 1989. We recognize that one of the most pervasive features of the new, multipolar security environment is its fluidity (which, in itself, has implications for a future export control system). However, we can say the following:

The Cold War is over.

Perestroika, the decline of ideology, the renunciation of the Brezhnev Doctrine, the collapse of communism in Central and Eastern Europe, the breaching of the Berlin Wall and the unification of

---

7. A June 1990 General Accounting Office report states that the US government is imposing an increasing number of foreign policy controls "to protect US security interests." This blurs the distinction between the two types of controls, and may harm US competitiveness, because foreign policy controls are easier to impose and harder to remove than national security controls. US General Accounting Office, *Export Controls: Commerce Has Improved Its Foreign Policy Reports to Congress*, GAO/NSIAD-90-169 (Washington, DC, 1990), pp. 3-5.

Germany, the withdrawal of Soviet troops from Central Europe, the disintegration of the Warsaw Treaty Organization,[8] and East-West cooperation in the 1990-1991 Gulf crisis are reasons and indicators that the ideological and military conflict that dominated East-West relations for 40 years is over. Clearly, the Soviet Union has neither the capacity nor the intention to launch a blitzkrieg on Western Europe,[9] or to continue a policy of radical activism in developing countries.[10] However:

The Soviet Union remains a military superpower.

The Soviet Union combines size and geostrategic reach with large and modern nuclear and conventional forces. It has over 11,000 nuclear warheads[11] and about 4 million active armed troops.[12] Moreover, US military experts point out that the Soviet military is intent on using perestroika to acquire Western technologies crucial to weapons development. One such expert, Sumner Benson, writes:

> Under the rubric of perestroika the Soviet Union has new, legal opportunities to obtain Western technology. . . . Mean-

---

8. See "Warsaw Pact Announced Force Reductions," in *Soviet Military Power 1989* (Washington, DC: US Department of Defense, 1989), p. 62; and comments of Secretary of Defense Richard Cheney, Lt. Gen. Harry Soyster of the Defense Intelligence Agency, Former Secretary of Defense James Schlesinger, former Defense Department official Richard Perle, and Gen. John Vessey, ret., former chairman of the Joint Chiefs of Staff, in Dale R. Herspring, "Reassessing the Warsaw Pact Threat: The East European Militaries," *Arms Control Today* (March 1990), p. 11.

9. "The threat of a cohesive Warsaw Pact, with strong Soviet forces based well forward in Central Germany, capable of attacking with a warning time of a few weeks, has disappeared." "NATO's Continuing Role in the 1990s," Report on presentation by Robert O'Neill, March 28, 1990, *The Woodrow Wilson Center Report* 2, no. 1 (June 1990), p. 1. Also see Herspring, "Reassessing the Warsaw Pact Threat."

10. See, for example, "Third World Cold War Ending," Report on presentation by Robert Litwak, January 22, 1990, *The Woodrow Wilson Center Report* 1, no. 4 (March 1990), pp. 3-4.

11. *SIPRI Yearbook 1989* (New York: Oxford University Press, 1989), p. 15. How much leverage these nuclear weapons provide is, of course, debatable.

12. Alexander Konovalov, "Counting Men, Here and There," *Moscow News*, March 11-18, 1990.

while, it has been widely recognized that illegal Soviet efforts to gain Western technology continue unabated, or have increased.[13]

While much in the Soviet Union has changed and continues to change as we enter the 1990s, no one will deny that it retains the military capabilities of a superpower. At a minimum, the United States would want to ensure that it prevented a Soviet first-strike nuclear capability and preserved its own second-strike capability.

Soviet and Central European reforms require Western economic and technological assistance.

The emerging democracies in the East are committed to replacing centralized planning with Western-style market economies. This is, undoubtedly, an enormous task. The Soviet and Central European economies are in desperate straits; Gorbachev and Central European leaders have called for Western economic assistance; and most economists (in the West and the East) believe that economic reform and revitalization will be more difficult without Western economic assistance, including dual-use high technology. Many also believe that economic revitalization and democratization are inextricably linked. Some have argued that certain technologies, for example, telecommunications and personal computers, promote democratic pluralism.[14]

Given the demise of the WTO and the revolutions in Central Europe, many now believe that the greatest potential security threat from Central Europe would be the failure of the economic and political

---

13. Sumner Benson, "The Security Perspective on Export Control Policy in the 1990s," in *Export Controls in Transition: Perspectives, Problems and Prospects*, ed. Gary K. Bertsch and Steven Elliott-Gower (Durham, NC: Duke University Press, forthcoming 1992), mimeo., pp. 29 and 30.

14. This argument is often heard and, of course, favored and promoted by those in the computer and telecommunications industries. For example: "If positive political change in Eastern Europe is to be sustained, it must be quickly buttressed by accessible, affordable, efficient and diverse means of communication." "Eastern European Communication Task Force Gains Momentum," *Eastern European & Soviet Telecom Report* 1, no. 1 (May 1990), pp. 1-2.

reforms taking place there, and the regional instability that would ensue.[15]

Security is a multifaceted concept; economic and technological competitiveness are vital to national security.

It is becoming increasingly clear that a Cold War conception of security that relates solely to military requirements is of limited utility. Michael Renner of the Worldwatch Institute writes:

> As more and more people will now agree, "national security" has for too long been defined in solely military terms. For too long we have measured security by counting the number of tanks and ships and planes and missiles on each side, and comparing the range, firepower, and position of our weapons with the comparable equipment of our enemy.[16]

Whereas Renner is concerned with environmental security, Inman and Burton (of the Council on Competitiveness) are concerned with economic security. They write that "national security can no longer be viewed in exclusively military terms; economic security and industrial competitiveness are also vital considerations."[17] Talking about the future of US export controls, industry representative Bill Chaskta put it this way:

> The EAR [Export Administration Regulations] goal has been to maintain the technological gap between the US and the Soviets, but national security must be redefined to involve a comprehensive competitiveness strategy in policy. The White

---

15. See William J. Long, "Defining Strategic Exports in the 1990s: From Export Control to the Management of Technology Exchange," in *Export Controls in Transition*, ed. Bertsch and Elliott-Gower, mimeo., p. 20.

16. Michael Renner, "From Military Security to Environmental Security: The Challenge of Trans-Boundary Environmental Perils," *PAWSS Perspectives* 1, no. 1 (April 1990), p. 9.

17. B.R. Inman and Daniel F. Burton, Jr., "Technology and Competitiveness: The New Policy Frontier," *Foreign Affairs* 69, no. 2 (Spring 1990), p. 133.

House and lead agencies must work with industry . . . in order to save the most critical strategic industries.[18]

Such arguments from business executives may be viewed by some as self-serving, but they have increasing merit and political support in the new security environment.

A crude indicator of the increasing importance of the economic component of national security is the growing contribution of foreign trade to US GNP, rising from less than 10% in the 1960s to almost 25% in the early 1990s.[19] An indication of the national security concern about technology and technological competitiveness is the plethora of studies in 1990 on "critical technologies," including the Commerce Department's *Emerging Technologies*,[20] the Defense Department's *Critical Technologies Plan*,[21] the Council on Competitiveness's *Strategic Assessment of National Technology Priorities and US Competitiveness*, and an Office of Science and Technology report on "national critical technologies." The Defense Department report identified 20 technologies that should be prime targets for federal research and development dollars because "they are critical to national security."[22] The Commerce Department report noted that "new technologies are the strongest assurance for maintaining a superior national security posture."[23]

---

18. "Comments of Bill Chaskta, Washington Resources International," *Conference Proceedings: Telecommunications, Computers and the Future of Export Controls*, Washington, DC, May 15-16, 1990 (Omaha, NE: International Center for Telecommunications Management, 1990), p. 26.

19. The Aspen Institute, *Economic Dynamism and Export Controls: International Technology Transfer in a Knowledge-Intensive World* (Queenstown, MD, 1988), p. ix.

20. US Department of Commerce, Technology Administration, *Emerging Technologies: A Survey of Technical and Economic Opportunities* (Washington, DC, 1990).

21. US Department of Defense, *Critical Technologies Plan*, Report for the US Congress, Committees on Armed Service, AD-A219 300, March 15, 1990.

22. "Pentagon Picks 20 Critical Technologies for Development," *Challenge* 3, no. 8 (April 1990), p. 1.

23. US Department of Commerce, *Emerging Technologies*, p. xv.

Changes in the concept of security are reflected in the views of average US citizens. In the 13th "Americans Talk Security" poll in 1990, nonmilitary threats (including foreign economic competition) were named 90% of the time when respondents chose the top two threats to national security, while military issues were named 51% of the time. Soviet nuclear weapons and Soviet aggression ranked 13th and 14th out of 14 listed threats.[24]

Certainly, protection from external military attack is a critical component of security, but there are also economic, technological, environmental, and other components that make up a *comprehensive* and *integrated* conception of security. These other components are becoming increasingly important to the whole, and some components (particularly US economic and technological competitiveness) are affected significantly by US export control policy.

US power in the world has decreased over the past 40 years.

Power is a relative concept. World War II left the United States in an "artificial" position of power and influence in an international order that was mostly of its own creation. But as Europe and Japan recovered from the devastation of the war and began to prosper, US power declined. Power and influence are not only relative but also amorphous concepts. (The current debate on the "decline" of the United States bears testimony to this fact.)[25] However, the United States has been having increasing difficulty asserting its will in international organizations, two relevant examples being COCOM and the European Bank for Reconstruction and Development (EBRD).

---

24. Americans Talk Security Project, "The Peace Dividend as the Public Sees It," Survey no. 13, Report no. 1 (Boston: ATS, 1990).

25. The "pessimists" include Paul Kennedy, *The Rise and Fall of the Great Powers* (New York: Random House, 1987); and Benjamin M. Friedman, *Day of Reckoning: The Consequences of American Economic Policy Under Reagan and After* (New York: Random House, 1988). The "optimists" include Richard Rosecrance, *America's Economic Resurgence: A Bold New Strategy* (New York: Harper & Row, 1990); Joseph Nye, *Bound to Lead: The Changing Nature of American Power* (New York: Basic Books, 1990); and Henry Nau, *The Myth of America's Decline: Leading the World Economy into the 1990s* (New York: Oxford University Press, 1990).

The West Europeans appear to have increased their power in COCOM since glasnost and perestroika came to the Soviet Union and reform and revolution swept through Central Europe in 1989. For a while, before the Defense Department moderated its hard line on export controls, the United States dragged its feet on "streamlining" COCOM's industrial control list. The "radical" proposals made by the US in early 1990 to reform COCOM were largely a more conservative restatement of European proposals, and arguably constituted an attempt to get back "out front" in COCOM. Many believe that, despite the rhetoric, the United States has failed to reassert its leadership in COCOM.[26]

The United States has also been unable to assert its leadership and influence in the establishment of the post-Cold War EBRD. The bank is a French initiative, led by a French citizen, and based in London. The United States initially opposed Soviet participation, but found itself isolated.[27]

These are not, of course, examples of the United States totally caving in to European pressure; there has been compromise in both COCOM and the EBRD. However, US influence in these organizations in the post-Cold War period seems considerably less than its influence in the creation and implementation of international organizations in the immediate post-World War II period. This diminution of influence has implications for the creation of any new post-Cold War export control regime.

In addition, the relative decrease in US political influence is

---

26. "West Falls Out Over Hi-tech Exports to the East," *The Times*, October 24, 1989; "It's 16 to 1 Against US on the Eve of Paris COCOM Meeting," *International Herald Tribune*, October 25, 1989; "US Eases Stance on Exports: Allies Urge Sales on Technology to Eastern Europe," *International Herald Tribune*, January 20, 1990. Obviously, newspaper headlines tend to overdramatize, but the trend they indicate here is an accurate one. See also Martin J. Hillenbrand, "Export Control Policy in the 1990s: The Diplomatic Perspective," in *Export Controls in Transition*, ed. Bertsch and Elliott-Gower, mimeo.

27. "US Threatens Not to Join Bank For East Europe if Soviets Benefit," *The New York Times*, March 15, 1990; "US Likely to Join Bank to Aid Reform in E. Europe," *Financial Times*, May 26, 1990; "A New Bank Plans East European Aid: 40 Nations Sign a Treaty for $12 Billion Institution—Soviets Get 6% Stake," *The New York Times*, May 30, 1990.

closely related to its decline in economic and technological preeminence.

International competition has eroded US technological superiority; there is a global diffusion of advanced technology.

The United States has lost its technological lead in numerous areas of advanced technology, including aircraft, radar, sonar, microcomputers, transformers, semiconductor manufacturing, radio, and television.[28] Some of these technologies are controlled by the United States and COCOM, but produced by non-COCOM countries, most notably the Asian newly industrialized countries (NICs). During the 1980s, for example, Taiwan became the world's sixth largest producer of computer hardware,[29] and South Korea became an important supplier of electronics and telecommunications equipment to the Soviet Union.[30] Moreover, as is pointed out below, there is a growing number of suppliers of other dual-use goods and technologies that may be used in the manufacture of long-range delivery systems and nuclear, chemical, and biological weapons.

There is a global proliferation of nuclear, chemical, and biological weapons, and missile technologies.

Many states (including "terrorist states" and those in unstable regions of the world such as the Middle East) have or are acquiring or

---

28. Numerous Department of Commerce "competitive assessments" attest to the decline in international competitiveness of some of these sectors. See, for example, US Department of Commerce, International Trade Administration, *A Competitive Assessment of the US Semiconductor Manufacturing Equipment Industry* (Washington, DC: US Government Printing Office, 1985).

29. "Taiwan and Korea: Two Paths to Prosperity," *The Economist*, July 14, 1990, p. 22.

30. Anticipating a relaxation of COCOM rules, Samsung signed an agreement with the Soviet Union to supply COCOM-controlled TDX telecommunications systems, incorporating 32-bit microprocessors. "Samsung Admits Soviet Deal Violates Cocom," Foreign Broadcast Information Service, *Daily Report--Eastern Europe* 90-115, June 14, 1990, p. 26.

developing weapons of mass destruction and long-range delivery systems. These include nuclear, chemical, and biological weapons, and ballistic missile systems.

There are a number of threshold nuclear weapons states throughout the world.[31] Many of them are not party to the Nuclear Non-Proliferation Treaty (NPT), and they have the capability to export certain nuclear weapons materials and technologies.[32]

According to CIA estimates, as many as 20 nations have or are developing chemical weapons capabilities, including Iraq, Iran, the PRC, North Korea, Taiwan, Burma, India, Pakistan, Syria, Israel, Egypt, Ethiopia, and Libya. The Soviet Union is suspected of supplying the Angolan government with chemical weapons, which were used against the UNITA guerrillas. Ten nations, including five in the Middle East, are suspected of developing viruses and bacteria for military use. Moreover, some subnational terrorist groups have threatened to use chemical and biological weapons.[33]

Countries in unstable regions are also acquiring the means to deliver these weapons of mass destruction over long distances. According to security analyst Janne Nolan:

> The proliferation of ballistic missiles and missile technology to the developing world is posing new challenges to US and global security, suggesting the potential for fundamental alterations in the international security environment in coming decades.

Nolan claims that the adaptation of technological resources to missile technology is taking place in at least ten countries, including Brazil, Argentina, India, Pakistan, Israel, North and South Korea, the PRC, Iran,

---

31. Leonard S. Spector, *The Undeclared Bomb* (Cambridge, MA: Ballinger, 1988), p. 18.

32. William C. Potter, *International Nuclear Trade and Nonproliferation: The Challenge of the Emerging Suppliers* (Lexington, MA: Lexington, 1990), p. 7.

33. "Chemical Weapons: Fighting to Stem the Tide," *Jane's Defence Weekly* Special Report, July 14, 1990, pp. 51-58. Also see Elisa D. Harris, "Chemical Weapons Proliferation in the Developing World," in *Defence Yearbook 1989* (London: Brassey's Defence Publishers, 1989), pp. 67-88.

and Iraq.[34] The CIA estimates that at least 15 nations will be producing their own ballistic missiles by the year 2000.[35]

Implications for US Policy

What does all of this mean for US export policy and any multilateral export control system to which the United States is a party?

The Cold War is over.

In the Summer 1990 issue of *Foreign Affairs*, Theodore Sorenson wrote:

> The touchstone for our nation's security concept—the containment of Soviet military and ideological power—is gone. The primary threat cited over forty years in justification for most of our military budget, bases and overseas assistance is gone.[36]

Many others believe that the threat or likely use of Soviet strategic capabilities is also gone. Soviet policy makers are consumed by domestic concerns; they show little interest in militarily confronting or challenging the West. And if, as many argue, East-West military conflict is gone, then so are the premise and raison d'être of Cold War export controls.

There are some who argue that the logical extension of this argument is to dismantle the Cold War system of export controls. The

---

34. Janne E. Nolan, "Ballistic Missiles in the Third World—The Limits of Nonproliferation," *Arms Control Today* (November 1989), pp. 9-14. Also see James Bruce, "Assessing Iraq's Missile Technology," *Jane's Defence Weekly*, December 23, 1989, pp. 1,371 and 1,374; and Pushpindar Singh, "India's Agni Success Poses New Problems," *Jane's Defence Weekly*, June 3, 1989, pp. 1,052-53.

35. "Controlling the Spread of Ballistic Missiles," *Jane's Defence Weekly*, April 22, 1989, p. 696.

36. Theodore C. Sorenson, "Rethinking National Security," *Foreign Affairs* 69, no. 3 (Summer 1990), p. 1.

June 1990 high-level meeting of COCOM resulted in significant decontrols, but as far as we know there was no discussion on dismantling COCOM. Rather, there was an agreement to, inter alia, create a "core list" of the most sensitive strategic goods and technologies[37] to replace the extensive industrial list. Skeptics might ask, if the Cold War is over and if containment is no more, then why is a core list necessary? Some would suggest that the answer has more to do with Cold War "mind-sets" and institutional self-preservation than with the realities of the new security environment.[38] And, as we know from our understanding of human and political behavior, mind-sets and institutions are usually slow to change.

Others believe that an important implication of the end of the Cold War is the increased possibilities of cooperation between East and West. An early and dramatic example of such cooperation was the Soviet Union's support of the Western position in the United Nations during the 1990-1991 Gulf crisis. Such support would have been almost inconceivable during the Cold War, and it indicates a new superpower willingness to move beyond traditional regional security interests to promote global security.

US-Central European cooperation on safeguards systems is another example of post-Cold War cooperation.[39] With further arms reductions, progress in the Soviet military-to-civilian conversion program, evidence that safeguards systems worked in Central Europe, and other confidence-building measures, US-Soviet cooperation in this and

---

37. The US government has divided these into eight broad categories: electronics design, development and production; advanced materials and materials processing; telecommunications; sensors, sensor systems and lasers; navigation and avionics systems; marine technology; computers; and propulsion systems. "Government and Industry Exerting Hard-Core Effort," *Export Control News* 4, no. 8 (August 15, 1990), p. 4.

38. As Igor E. Artemiev puts it in this volume: "the Western perception of the [Soviet] military-political threat has not yet evaporated from the minds of decision makers. . . .[who] are not yet convinced that new political and economic realities prevail or that COCOM and export controls are irrelevant or at least obsolete."

39. "US, Central Europe Cooperate on Safeguards," *Export Control News* 4, no. 8 (August 15, 1990), pp. 3-4. Also see Andrzej Rudka's contribution to this volume.

other areas of technology transfer and technology control would be possible.[40]

The Soviet Union remains a military superpower.

An essential difference between the Central European countries and the Soviet Union remains, however, and this involves the Soviet Union's capacity to inflict unacceptable damage on the United States and its Western allies. When combined with 40 years of suspicion and mutual mistrust, the Soviets' military might means that, despite Soviet willingness to permit on-site end-use verification of Western high-technology imports, more confidence-building measures of the sort mentioned above will be necessary before US and Soviet policy makers can sit down to negotiate safeguards agreements.

One must be sensitive to Western and particularly US attitudes towards the Soviet Union. Although Gorbachev and his reformers have softened US distrust and fear of his country, there are still likely to be sizable US constituencies in the future who will for one reason or another oppose East-West technological cooperation. One can imagine members of the US Congress gaining considerable political advantage among their local constituencies by revealing that the United States was "selling the rope," regardless of whether it was to communist or non-communist end-users in the Soviet Union.

Soviet and Central European reforms require Western economic and technological assistance.

In June 1990, COCOM members agreed to reduce technological controls to the Soviet Union and Central Europe, and permit more favorable treatment to those Central European countries that instituted acceptable technology diversion safeguards regimes (their access to advanced Western dual-use technology would be limited only by the core list, and of course their purchasing power). The decision reflected

---

40. For a discussion of a verifiable safeguards systems in the Soviet Union, see Artemiev's chapter in this volume.

the allies' belief in 1) the importance (whether real or symbolic) of Western technology to Central European reform; 2) the reduced military threat of the WTO; and 3) the continued military threat posed by the Soviet Union.

To transfer technology to Central Europe while controlling it to the Soviet Union, US officials began working with their counterparts in Poland, the CSFR, and Hungary to create technology diversion safeguards regimes.[41] COCOM members apparently decided that the implementation of such safeguards in these countries would justify the approval of truly cutting-edge technology exports, and yet avoid what has been called the US fear of a "technology bazaar or a free technology zone" in Central Europe.[42]

William Long of the Georgia Institute of Technology argues that the economic and political reforms in the Soviet Union and Central Europe call for more than a change in policy; they call for a reconceptualization of security strategies for the 1990s. He writes:

> Blocking the East's access to technology will no longer be the overriding security strategy for the United States and the West. The changed political landscape opens the possibility that technology transfers which foster democracy or market reforms in the East may prove a more effective strategy in enhancing security than a policy of technology denial.[43]

Thus, there is an increasing realization that the West may need to *transfer* rather than *control* technology to the East for strategic reasons. In this context, "strategic" technologies are reconceptualized to mean those technologies, such as computer and telecommunications technologies, that can assist the processes of economic and political

---

41. "US, Central Europe Cooperate on Safeguards;" and "Interview: Bob Price, Director, Office of COCOM Affairs, Department of State," *Export Control News* 4, no. 9 (September 21, 1990), pp. 13-15.

42. "US, Central Europe Cooperate on Safeguards," p. 3.

43. Long, "Defining Strategic Exports in the 1990s."

reform in the East. To take this line a step further again means dismantling COCOM, an organization that is designed to stifle technology transfer and that, with this reconceptualization, thwarts Western strategic objectives. A spokesperson at the West German embassy indicated this frustration when he said: "In an age where via dialogue and cooperation we try to assist reform processes in Poland, Hungary and the USSR, COCOM is outdated."[44]

For the present, however, Western (and especially US) policy makers continue to believe that, as far as the Soviet Union is concerned, the strategic benefits of controlling technology still outweigh those of transferring it. Just how long this will be the case is unclear. It may be a long time, if Western policy makers are basing differentiation on military superpower status and/or political instability. By virtue of its turbulent history and great size, the Soviet Union is likely to retain a large standing army for the foreseeable future. Despite reductions, it is also likely to continue to produce and deploy some modern nuclear weapons. And with all its tremendous domestic political problems (especially the nationalities problem), the Soviet Union will remain politically unstable. Of course, while the security costs might remain stable, the security benefits of transferring controlled technology to the Soviet Union may increase (as a result of the Central European experience), and supersede those costs. Also, the United States may well face increased pressure from the European allies to discard the policy of differentiation, especially given progress on arms reductions, the military-industrial conversion program, and evidence that verifiable safeguards systems are working in Central Europe.

Of course, all of this assumes that Western high technology can assist the processes of economic and political reform in the Soviet Union and Central Europe. As Volkhart Vincentz points out in this volume, there is a considerable body of literature on the economic impact of high-technology imports, but it is far from conclusive. The question of whether and to what extent 1990s technology can be applied in and benefit these economies is worthy of further study.

---

44. "A Debate on Controls of Exports," *The New York Times*, October 9, 1989.

Security is a multifaceted concept; economic and technological competitiveness are vital to national security.

Although little had been done about it until recently, the economic costs of export controls have long been recognized. According to some estimates, national security and foreign policy controls cost US exporters some $12.5 billion a year in lost sales during the early 1980s.[45]

It is becoming increasingly clear to many US policy makers that unilateral US controls or multilateral controls with leaks (often resulting from less than vigorous implementation by some of its technology producing and exporting allies) can come at considerable cost to the US economy. The United States may become less inclined in the future to impose these controls on items that will be exported or re-exported by its economic competitors. In sum, there is evidence leading some to conclude that the United States is becoming more sensitive to the relationship of export controls to its own economic and technological competitiveness, and to the relationship of all of this to its economic (and hence national) security.[46]

US power in the world has decreased over the past 40 years. International competition has eroded US technological superiority; there is a global diffusion of advanced technology.

The implications of these two fundamentals are closely related. Experience shows us that any effective effort to control the transfer of

---

45. See Committee on Science, Engineering, and Public Policy of the National Academy of Sciences, the National Academy of Engineering, and the Institute of Medicine, *Balancing the National Interest: US National Security Export Controls and Global Economic Competition* (Washington, DC: National Academy Press, 1987), pp. 114, 121, and 246; President's Commission on Industrial Competitiveness, *Global Competition: The New Reality*, vol. 2 (Washington, DC: US Government Printing Office, 1985), pp. 194 and 197; and Stanley D. Nollen, "Business Costs and Business Policy for Export Controls," *Journal of International Business Studies* 18, no. 1 (Spring 1987), p. 8.

46. See, for example, Committee on Science, Engineering, and Policy, *Balancing the National Interest*; The Aspen Institute, *Economic Dynamism and Export Controls*; and The Business Roundtable, *Focus on the Future: A Global Export Control Framework* (Washington, DC, 1990).

"dangerous" technologies must be multilateral, and must include all the major suppliers of the controlled technologies.

This was less of an issue, of course, when the United States was the preeminent producer of those technologies it sought to control. But the (relative) economic and technological decline of the US and the emergence of numerous, new, widely available dangerous technologies have made the export control enterprise more difficult in a number of respects. First, the United States has had to work to incorporate new exporters of controlled technologies into the COCOM regime, with limited success. Second, the proliferation of advanced technologies and the emergence of new dual-use technologies to be controlled means that there is a vast number of suppliers and potential suppliers who need to cooperate. Third, because the United States has generally used uninhibited access to US technology as an incentive for other countries' cooperation, the decline of US technological superiority makes that cooperation more difficult to obtain. The mixed success of the Third Country or 5(k) Initiative, which seeks COCOM-like controls in non-COCOM countries (especially the NICs), is evidence of this.[47] Of course, on the other hand, the willingness of the Soviet Union and Central Europe to adopt safeguards systems suggests that access to US (*but also* Western) technology is sufficient incentive for them.

The general point remains: an effective export control system must be multilateral, and the benefits of belonging must clearly outweigh the costs. The United States cannot create sufficient incentives on its own any more. The creation of incentives, the building of benefits, also has to be a multilateral effort. All of this will be increasingly difficult in the 1990s.

There is a global proliferation of nuclear, chemical, and biological weapons, and missile technologies.

As the communist threat has diminished, the threat posed by the proliferation of weapons of mass destruction has risen higher on the political agenda, and has captured the attention of export control

---

47. Some have argued that access to the US domestic market would be a much more potent tool to encourage cooperation on export controls.

policy makers in the United States and abroad.[48] This gives new meaning and purpose to the export control enterprise. The task is to examine whether and how export controls can prevent or slow this proliferation.

The problems associated with controlling the materials and technologies associated with these weapons are familiar. First, there is the dual-use problem. What should be controlled? Many chemicals that could be used in a weapon have legitimate civilian uses. Indeed, some contend that because chemical weapons are generally made by combining relatively harmless chemical agents—precursors—it is almost impossible to control trade to prevent chemical weapons manufacture.[49] Ballistic missile technology is also dual-use. The United States believes that civilian space programs have been used as conduits for launch and guidance systems for military ballistic missiles. The United States has consequently restricted the transfer of such technology to countries that are developing nuclear-capable missiles. The European partners of the US in the Missile Technology Control Regime (MTCR), particularly France, have decided to promote what they consider legitimate space programs, and do not draw a connection to military ballistic missile programs.[50]

Even if agreement can be reached on what to control, who should it be controlled from? During the Cold War, the communist targets of export controls were clear. The criteria for identifying target states will be less clear and probably more fluid in the post-Cold War period. Which countries have "legitimate" space programs and which are more interested in having military ballistic missiles? How does one classify terrorist states? The distinction between a terrorist and a "freedom fighter" may be a very fine one. A post-Cold War export control system will need to be flexible and adaptive. Of course, when push came to shove, COCOM proved to be flexible, with the PRC in

---

48. A State Department official was reported to have said that the United States, its allies, and several neutral nations were beginning "to redefine their national security in terms of preventing hostile Third World nations from securing access to weapons of mass destruction." "Interview: Bob Price," p. 15.

49. "Chemical Weapons: Fighting to Stem the Tide," p. 58.

50. US General Accounting Office, *Arms Control*, p. 17.

1957 and again in 1985, and with certain Central European countries in 1990. But then, much more flexibility will be required in the future.

An export control regime designed to prevent this proliferation will require far broader membership than COCOM. It will need to include the NICs and many of the former targets of export controls. For example, to be effective, the MTCR will need to include other suppliers of missile technology, most importantly the Soviet Union and the PRC.[51] The end of the Cold War makes such East-West cooperation on export controls possible, but under what conditions?

As was mentioned earlier, broad participation in a comprehensive export control regime will require the leaders or current members of such a regime to create a structure of incentives to join and stay, and disincentives for defections. Access to high technology and certain financial tools could be used in such a structure.[52]

Conclusion

The new security fundamentals of the post-Cold War world suggest in broad outline what a new export control regime might look like. Such a regime would probably encourage as well as control the transfer of strategic technologies. Concerning controls, there would be more emphasis on munitions and dual-use technologies that enhance nuclear, chemical, and biological weapons and long-range missile systems. A new export control regime would probably need to be much larger than COCOM, incorporating both the NICs and the

---

51. The Soviet Union is a major arms exporter. See, for example, "Growing Soviet Export: Military Technology," *The New York Times*, May 3, 1990.

52. According to a State Department official, "there have been discussions with these countries [Poland, Hungary, and the CSFR] under the aegis of the multilateral regimes for controls of missile, nuclear and chemical/biological weapons." The official was optimistic about the prospects for cooperation on these types of controls. "Interview: Bob Price," p. 14. However, others are concerned that "cash-starved" companies in Central Europe will be unable to resist the temptation to export and re-export sensitive technologies to profitable Middle Eastern markets. "Easing of Technology Export Controls May Boost Arms Smuggling in Mideast," *The Wall Street Journal*, June 19, 1990.

West's former communist adversaries. It would thus be based on global cooperation rather than East-West conflict. Finally, controls would be targeted at terrorist states and those states believed to be acquiring weapons of mass destruction; broadly speaking, a shift from communist and post-communist countries to certain developing countries and NICs who are unwilling to participate in an export control regime.

Even assuming that the new security fundamentals and their export control implications are accepted, and assuming there is some consensus on what a new export control regime should look like, a multiplicity of questions remain unanswered, not the least of which concern the institutional framework of such a regime, and the growing difficulties of technology controls in the modern world.

Some people have argued that the mandate and membership of COCOM should change to allow it to coordinate a new post-Cold War export control system. They argue that despite its informal nature, COCOM is the only organization with a proven track record in export controls,[53] and at the very least it can provide important operational lessons about multilateral export controls.[54] On the other hand, of course, COCOM's experience is in controlling and not transferring technology. (Whether there needs to be an active institutional mechanism of technology transfer or whether this can be left to the market is another matter.) There is also the possibility that COCOM's close association with NATO and its ideological baggage may not make it a suitable forum for East-West cooperation. Although this seems like a plausible concern, the Soviets' and Central Europeans' desire for Western technology is apparently so great that they may be prepared to forget their scathing criticisms of COCOM in the past, and begin to work in a cooperative mode in a reformed COCOM. Because it includes both NATO and WTO countries and has security and

---

53. See, for example, Lionel H. Olmer, "COCOM Adapts Strategic Trade Rules," *The Atlantic Council Bulletin* 1, no. 17 (July 19, 1990).

54. Nolan, "Ballistic Missiles in the Third World," p. 12. Also see Peter Rudolf, "Beyond Containment: COCOM Needs Change From the Ground Up," *International Herald Tribune*, June 6, 1990; and "Cocom Considered as Arms-Sale Curb: US, W. Germany Want Group to Police Third World Proliferation," *The Washington Post*, August 2, 1990.

economic "baskets," an institutionalized Conference on Security and Cooperation in Europe (CSCE) has been discussed as an alternative successor to COCOM.[55] Yet proliferation is a global problem that may require a global rather than a regional solution. If this is the case, perhaps the Nuclear Non-Proliferation Treaty provides a model for a future technology transfer/control regime that might be coordinated under the auspices of the United Nations. Certainly, the NPT combines the elements of control and transfer that a new technology management regime may need; and the UN's recent experience with controls on Iraq may provide valuable experience that will help to inform us of the feasibility of this alternative.

Other difficult issues remain. One concerns the very efficacy of technology controls in the 1990s. When the flow of illegal drugs cannot be controlled, when millions of people move freely between East and West (and North and South), and when militarily sensitive technology can be miniaturized, posted on a computer disk, or even telefaxed, doubt is cast over the entire export control enterprise.

---

55. Michael R. Lucas, "The Abolition of COCOM and the Establishment of a Technology Disarmament and Transfer Agency in the CSCE," *Bulletin of Peace Proposals* 21, no. 2 (1990), pp. 219-25. And, more generally, "A New Europe Ponders Security," *The Christian Science Monitor*, April 16, 1990. Lucas's proposal warrants careful consideration.

COLLOQUIUM PARTICIPANTS

"Technology Transfer in the 1990s"
Institute for East-West Security Studies, New York
September 14, 1990

Gary K. Bertsch
Center for East-West Trade
 Policy
University of Georgia
Athens, Georgia

Wolfgang Bruckmann
IEWSS Resident Fellow
New York

Eva Culen
The Conference Board
New York

Steven Elliott-Gower
Center for East-West Trade
 Policy
University of Georgia
Athens, Georgia

Larry W. Evans
Patent & License Division
BP America
Cleveland

Janos Farkas
Hungarian Commercial Office
New York

Lothar Griessbach
German Consortium of Industry
Washington, DC

Leah Haus
Politics Department
New York University
New York

George Holliday
Congressional Research Service
Library of Congress
Washington, DC

Catherine Jewsbury
WEFA Group
Washington, DC

Karen Kalina
CDS International Inc.
New York

Sergei L. Kambalov
Department of International
 Economic and Social Affairs
United Nations
New York

Barbara Katz
Stern Graduate School of
 Economics
New York University
New York

David M. Kemme
IEWSS
New York

Colloquium Participants

Elina Kirichenko
Institute of World Economy and
  International Relations
  (IMEMO)
Moscow

Marie Lavigne
IEWSS Resident Fellow
New York

William J. Long
School of International Service
American University
Washington, DC

Reneo Lukic
IEWSS Resident Fellow
New York

Karl M. Mayer-Wittmann
Mayer-Wittman Joint Ventures
Old Greenwich, CT

Alexei Mozhin
IEWSS Resident Fellow
New York

John Edwin Mroz
IEWSS
New York

Itaru Murata
Consulate General of Japan
New York

Henry Nau
George Washington University
Elliott School of International
  Affairs
Washington, DC

Nikolaj Ordnung
Developing Countries Division
Institute of Economics
Czechoslovak Academy of
  Sciences
Prague

Andrzej Rudka
Foreign Trade Research Institute
Warsaw

Larry Schein
The Conference Board
New York

Paul Schroeder
National Committee on United
  States-China Relations
New York

Jozef van Brabant
Centrally Planned Economies
  Division
Department of International
  Economic and Social Affairs
United Nations
New York

Peter van Ham
Leiden University
Leiden

Volkhart Vincentz
Osteuropa-Institut
Munich

Peter M.E. Volten
IEWSS
New York

Simon Winder
*Reference and Scholarly
    Department
St. Martin's Press
New York*

Sergei Zelenev
*Development Analysis Section
Department of International
    Economic and Social Affairs
United Nations
New York*

IEWSS Staff
Kristi Bahrenburg
Tiffany Barnard
Marcia Dam
Carol Derfner
Claire Gordon
Barbara Hachmann
Rosalie Morales Kearns
Lesley Lyons
Rosemarie Roqué
Keith Wind

ABOUT THE AUTHORS

Igor E. Artemiev is Head of the Trade Policy Section of the Institute of World Economy and International Relations (IMEMO) in Moscow. He previously served as the Economic Affairs Officer with the UNCTAD Secretariat in Geneva. Dr. Artemiev's research interests include international trade and technology transfer. His recent publications include *Technology Markets in the World Economy* (Nauka: Moscow, forthcoming); *External Economic Relations and Technological Change* (Nauka: Moscow, 1989); "GATT and the USSR's Interests," *Mirovaia Ekonomika i Mezhdunarodnye Otnoshehiia,* no. 8, 1989 (with S. Stankovsky); and "The USA and the Uruguay Round of GATT," *SShA: Ekonomika, Politika, Ideologiia,* no. 1, 1988.

Gary K. Bertsch is Co-Director of the Center for East-West Trade Policy and Professor of Political Science at The University of Georgia. His research focuses on the politics of US and Western trade policy toward the Soviet Union and Eastern Europe. He teaches courses on communist and post-communist political systems and East-West trade policy. His publications include *After the Revolutions: East-West Trade and Technology Transfer in the 1990s* (Boulder, CO: Westview, 1991) (coeditor and contributor); *Reform and Revolution in Communist Systems* (New York: Macmillan, 1991); and *Controlling East-West Trade and Technology Transfer: Power, Politics and Policies* (Durham, NC: Duke University Press, 1988) (editor and contributor). Professor Bertsch is active in several projects intended to contribute to new East-West trade and economic policies for the 1990s.

Steven Elliott-Gower is Assistant Director of the Center for East-West Trade Policy and Adjunct Assistant Professor in the Department of Political Science at the University of Georgia. He is the author of numerous papers on East-West trade and export control issues, and is the co-editor of *Export Controls in Transition: Perspectives, Problems and Prospects* (Durham, NC: Duke University Press, forthcoming 1992); and *The Impact of Governments on East-West Economic Relations* (New York: New York University Press, 1991). His current research focuses on the concept of security and the future of export controls, a major project of the Center for East-West Trade Policy.

About the Authors

David M. Kemme is Pew Economics Scholar-in-Residence at the Institute for East-West Security Studies in New York. He was a Fulbright Lecturer at the Main School of Planning and Statistics in Warsaw (1981–1982), and has researched and published extensively on productivity and efficiency in Eastern Europe and the Soviet Union, disequilibrium macroeconomics, and the chronic shortage model of centrally planned economies. He has published in scholarly journals such as *Soviet Studies, Journal of Comparative Economics,* and *Jahrbuch der Wirtschaft Osteuropas,* and most recently has authored *Economic Transition in Eastern Europe and the Soviet Union: Issues and Strategies,* IEWSS Occasional Paper Series, no. 20 (New York, 1991) and edited *Economic Reform in Poland* (Greenwich, CT: JAI Press, 1991). In addition, Dr. Kemme is North American Editor of *Economics of Planning.*

Andrzej Rudka from the Foreign Trade Research Institute in Warsaw is currently Senior Fulbright Scholar at the Department of Economics and the Elliott School of International Affairs, George Washington University, Washington, DC. In 1990 he was a Visiting Fellow at the Institute for East-West Security Studies in New York; in 1989 he was a Visiting Professor at the College of Saint Rose in Albany, New York. He is a permanent member of the Working Group on Medium-Term Prospects and Structural Changes in the World Economy of the European Association of Business Cycle Institutes. Dr. Rudka is the author of numerous articles and publications on East-West trade, technology transfer and export controls, Polish-US economic relations, the US and the world economy, and Polish foreign trade.

Volkhart Vincentz is a senior member of the research staff of the Osteuropa-Institut in Munich. He has written several reports for the German Federal Ministry of Economics and other institutions on the economic development of the Soviet Union, production and trade of technology, and problems of economic reforms in Eastern Europe and the USSR. He is the author of numerous articles on utilization, diffusion, and high-technology trade in the East.

INDEX

Adler-Karlsson, Gunnar, 25
advanced materials, 11, 31
Afghanistan, 8, 10, 84
Africa, 27
aircraft, 29, 64, 68, 115
"Americans Talk Security" poll, 113
Angola, 116
Argentina, 116
arms control, 15
Australia, 90, 107n
Austria, 90
author's certificate, 14, 80–81, 88

Baltic republics, 11
banking systems, 23, 61, 70
Bauer, Tamas, 96
Benson, Sumner, 109
Berlin Wall, 5, 49, 103, 108
bikini list, 51
biological weapons, 107, 115–16, 123–24, 125
biotechnology, 75
Boeing, 29
Brada, Josef, 4
Brazil, 116
Burton, Daniel F., 54, 111
Bush administration, 10, 11, 30, 97

Cambodia, 107
Carter administration, 8, 10
Central Intelligence Agency (CIA), 107, 116, 117
Charter of the International Atomic Energy Agency, 99
Chastka, Bill, 111
chemical weapons, 107, 115–16, 123–24, 125

China, People's Republic of, 5, 11, 21, 23, 54, 55–56, 61, 106, 116, 124–25
command economies (centrally planned; socialist; state-owned), 4, 15, 58, 73, 74, 80–81, 85–86, 91, 96, 102, 110
Commodity Control List, 22, 91
Comprehensive Program of Scientific and Technological Progress, 84
Comprehensive Program of Socialist Economic Integration, 84
computers, 10, 11, 13, 23, 31, 52, 55, 59–60, 61, 67, 97, 110, 115, 120, 127
Conference on Economic Cooperation in Europe, 96
Conference on Security and Cooperation in Europe (CSCE), 16, 35, 74, 96, 127
confidence-building measures, 15, 96, 103, 119
consumer electronics, 55
Coordinating Committee for Multilateral Export Controls (COCOM), 1, 3, 5–8, 9–22, 25–26, 29–33, 35, 38–40, 41–46, 49–53, 55–56, 58, 61–62, 66, 90–91, 93–94, 96–98, 100–101, 103, 106, 108, 113–15, 118–21, 123–27
  Atomic Energy List, 6, 21, 106
  International Industrial List, 6, 21–22, 106, 118
  Munitions List, 6, 21, 106
core list, 10–11, 15, 31, 97, 103, 118–19

Council for Mutual Economic Assistance (CMEA), 3, 14, 55–56, 59–60, 63, 74, 81–86
Council on Competitiveness, 54, 111, 112
covert sales, 8
Cuba, 7, 107
customs-free zones, 45–46
Czech and Slovak Federal Republic (CSFR), 3, 11, 29, 37, 38, 39, 52, 53, 120

Decree on Workers' Control, 80
democratization, 12, 29, 48, 50, 53, 61, 95, 110, 120
detente, 2, 7
deterrence, 3, 12
Deutschlandpolitik, 49
dual-use technology, 6, 21, 51, 52, 65, 95, 100, 106–107, 115, 124
dumping, 69

Eastern Europe, 1–4, 10–17, 20–21, 23–24, 26, 29–35, 37–40, 42–43, 45–55, 57–63, 65–67, 69–70, 73, 77, 79, 81–82, 85–86, 91, 93–94, 96–97, 102–103, 106, 108, 110, 114, 118–21, 123, 125–26
Economic Commission for Europe (ECE), 35, 96
Egypt, 116
electronics, 11, 31, 59–60, 64, 70
environment, 14, 61, 70, 113
Ethiopia, 116
EUREKA, 28n, 38n, 66
European Bank for Reconstruction and Development (EBRD), 113–14
European Community (EC), 28, 63, 78

European Payments Union, 69
"exception request" principle, 22, 39, 55–56, 106
Export Administration Regulations (EAR), 107, 111
export controls
  costs to West of, 16, 24, 47, 53, 95, 121, 122
  effectiveness of, 4, 23–24, 25, 27, 28, 35, 50–51, 52–54, 58, 60, 70, 91
  for foreign policy reasons, 1–2, 7, 12, 90–92, 108, 111
  rationales for, 1–2, 4, 8, 74, 90–91, 93, 96, 105–106, 108
export denial, 1, 2, 44–45, 120

financial services, 35
Finland, 90
foreign trade organizations (FTOs), 83, 87–88, 89
France, 5, 8, 18, 22, 26, 114, 124
full employment guarantees, 4, 58

gas, 8–9
gas pipeline dispute, 9
General Agreement on Tariffs and Trade (GATT), 2, 20, 35
  Uruguay Round, 37
German Democratic Republic (GDR), 30, 41, 49–53, 59–61, 70
Germany, Federal Republic of (FRG), 1, 26, 49–50, 53, 61, 74, 108–109, 121
glasnost, 114
Gorbachev, Mikhail, 51, 66, 110, 119
grain, 62
Greece, 26
Gulf crisis, 109, 118

hard currency, 62
high-definition television (HDTV), 67
high-technology trade, 4, 6, 8, 10, 13–14, 16, 28, 37, 41, 54–59, 61–63, 66, 68, 70, 89–90, 95, 106, 119
human rights, 8, 107
Hungary, 3, 11, 29–30, 37, 38, 39, 52, 53, 85, 98, 120–21

import certification and delivery verification (IC/DV), 98–99
India, 116
Inman, B.R., 54, 111
Institute for International Economics, 25
Intermediate Nuclear Forces (INF) Treaty, 29, 99
International Business Machines (IBM), 67
international law, 19–20
interventionism, 14, 68
intra-industry trade, 63–64, 69, 76
Iran, 107, 116
Iraq, 107, 116–17
Israel, 116
Italy, 18, 26

Japan, 22, 27, 28, 42, 77, 78, 113
Jewish emigration, 11
joint ventures, 14, 36, 62, 86, 89

Kohl, Helmut, 74, 93
Korean War, 5

lasers, 11, 31
less developed countries (LDCs), 57, 60, 82, 94
Libya, 107, 116
licensed technology, 9, 36, 92
locomotives, 58–59

Long, William, 120

Macdonald, Stuart, 95
machine tools, 11, 52, 59, 97
management techniques, 36
marine technology, 11, 31
market economies, 14, 29, 35, 42, 47, 51, 53, 65, 70, 73–74, 96
McKinsey & Co., 66
mercantilism, 69
military-industrial complex, 96, 98
Missile Technology Control Regime (MTCR), 124–25
multinational corporations (MNCs), 14, 27, 76, 91

National Policy Review of Technological Progress in the USSR, 75, 90
national security, changing perceptions of, 15, 34, 53, 54, 73–74, 94, 105, 108, 113–14, 117, 120–21
Nau, Henry, 3
navigation and avionics systems, 11
newly industrialized countries (NICs), 16, 27, 57, 65, 94, 115, 123, 125–26
Nolan, Janne, 116
"non-strategic trade," 91
North Atlantic Treaty Organization (NATO), 41, 100, 108, 126
North Korea, 107, 116
Nuclear Non-Proliferation Treaty (NPT), 16, 116, 127
nuclear power, 61
Nuclear Regulatory Commission, 107

Office of Science and Technology, 112

oil, 8–9
Organization for Economic Cooperation and Development (OECD), 36, 77
Ostpolitik, 49–50

Pakistan, 116
patents, 14, 36, 80, 88
perestroika, 73, 103, 108, 109, 114
Poland, 1, 3, 7–9, 11, 23, 29–30, 37–39, 47, 52–53, 62, 84, 98, 120–21
pollution, 61
propulsion systems, 11

Reagan administration, 9–10, 91
re-exports, control of, 2, 43, 44–46, 48, 98–99
Renner, Michael, 111
research and development (R&D), 14, 64–66, 71, 76, 80, 82–83, 85, 88
Romania, 30
Russian Revolution, 80

safeguards systems, 13, 44–47, 52, 121
sanctions, 2, 3, 44, 50
semiconductors, 60, 115
sensors, 11, 31
Shcharansky, Anatolii, 8
Shevardnadze, Eduard, 97, 101
Silicon Valley, 66
Singapore, 90
Sofia principles, 81–82
Sorenson, Theodore, 117
South Africa, 107
South America, 27
South Korea, 90, 115, 116
Soviet Union, 1–5, 7–12, 14–16, 21, 31–32, 39–42, 49–54, 58–63, 65, 73–75, 77–80, 82–83, 85, 87–94, 96–102, 107, 109–11, 113–21, 123, 125–26
Foreign Trade Decentralization Order No. 203, 92
Licensintorg, 87–8
Ministry of Foreign Trade, 87, 92
Supreme Soviet, 88, 92
Spain, 90
standardization, 66–67
State Committee for Science and Technology (GKNT), 93
State Patent Fund, 88
steel, 58, 92
Strategic Arms Limitations Talks (START), 29
strategic goods, definition of, 2–3, 9, 12, 20–21, 28
Sweden, 77, 90
Switzerland, 77, 90
Syria, 107, 116

Taiwan, 115–16
Technical Advisory Committee, 7
technology, disembodied, 14, 77–79, 88, 92–93, 100–101
telecommunications equipment, 3, 11, 14, 23, 31, 35–36, 52, 55, 59–60, 62, 64, 66, 97, 110, 120
Third Country Initiative, 123
tourism, 35

United Kingdom (UK), 1, 5, 18, 22, 51, 77, 95
United Nations (UN), 16, 20, 118, 127
United Nations Conference on Trade and Development (UNCTAD), 35
United States (US), 1–2, 4–10, 15, 17–20, 22, 24, 26–30, 32–33, 36, 38–43, 45–46, 54–55, 57,

66, 77–79, 90–91, 93, 98–99
105–108, 110–14, 116–17,
119, 121–23
Arms Export Control Act (1976),
106
Atomic Energy Act (1954), 106
Commerce Department, 40, 107,
112
Congress, 11, 40–41, 119
Customs Service, 40
Defense Department, 10, 40, 91,
107, 112, 114
Energy Department, 107
Export Administration Act
(1969), 7
Export Administration Act
(1979), 8, 41, 106
Export Control Act (1949), 5, 7
Jackson-Vanik Amendment
(1974), 11
Omnibus Trade and Competi-
tiveness Act (1988), 10
State Department, 40, 107
Trading With the Enemy Act
(1917), 5
Treasury, 40

verification, 15, 40, 44–45, 97–
103, 119
Vietnam, 107

Warsaw Treaty Organization
(WTO), 1, 3, 7, 29, 41, 108–
10, 120, 126
World Bank, 62
World War II, 5, 28, 69, 106,
113–14
Worldwatch Institute, 111

Yemen, People's Democratic Republic of, 107
Yugoslavia, 82